# Geology of the San Juan Islands

NED BROWN

West Shore of Lummi Island

Copyright © 2014 by Chuckanut Editions

All rights reserved. No part of this publication may be reproduced, distributed, or transmitted in any form without the prior written permission of the publisher, except in the case of brief quotations embodied in critical reviews.

ISBN 9780989289139

LCCN 2014953626

Printed in Bellingham WA, USA.

Village Books
1200 11th Street
Bellingham WA 98225
360 671-2626

1st edition.

# CONTENTS

INTRODUCTION     *5*

ACKNOWLEDGEMENTS     *7*

PART I – BACKGROUND GEOLOGY

    Chapter 1. Tectonic Concepts     *9*

    Chapter 2. Geologic Time and Dating     *13*

    Chapter 3. Rock Terminology and Origins     *16*

PART II – GEOLOGIC HISTORY OF THE SAN JUAN ISLANDS

    Chapter 4. Overview of the San Juan Islands     *26*

    Chapter 5. Turtleback Complex     *32*

    Chapter 6. East Sound Group     *35*

    Chapter 7. Orcas Chert and Deadman Bay Volcanics     *39*

    Chapter 8. Ocean Floor Assemblage     *43*

    Chapter 9. Fidalgo Ophiolite     *51*

    Chapter 10. Nanaimo Group     *58*

    Chapter 11. Chuckanut Formation     *68*

    Chapter 12. Terrane Travels     *73*

    Chapter 13. Continental Glaciation     *79*

PART III – FIELD GUIDE

    Chapter 14. Good Places to See Rocks in the San Juan Islands     *87*

GLOSSARY     *94*

REFERENCES     *98*

*Cover photo:* fold in bedded greywacke, Eliza Island

*Kingfisher, Inati Bay, Lummi Island*

# INTRODUCTION

The San Juan Islands of northwest Washington State lie in a broad inland waterway between Vancouver Island and the mainland (Figs. 1,2). The Islands attract geologists both because they record dynamic earth processes that have grown the continent at its western margin, and because in comparison to the mountainous terrain of the mainland the Islands offer much easier access and working conditions and excellent water-washed shoreline exposures. Geologic interest in the Islands comes also from the general public, including Island residents and the many visitors that arrive from nearby population centers to enjoy the rural landscapes, coastal areas and recreational waterways.

Geologic study is a trip through time. Rocks tell a story that is unraveled by deducing how they were formed, and by dating the rocks using fossils and radiogenic decay in minerals and organic material.

These observations give us a chronicle of earth evolution that for the San Juan Islands is amazing in age, variety, and scale of events. Geologic processes affecting the Islands over the past 500 million years include: large-scale plate tectonic displacements, continental growth, volcanism, deep crustal burial and exhumation, and continental glaciation. The geology on display here invites an examination and understanding that will take the interested layman far into knowledge of how the earth works. In

**Fig. 1 Kayak geology at the north end of Blakely Island, with view east to Lummi and Cypress islands and Mt Baker.**

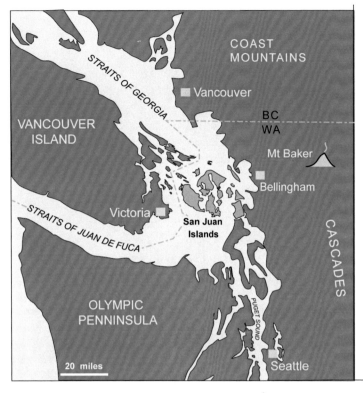

**Fig. 2 Geographic locale of the San Juan Islands, in the inland waterway flanked by the Straits of Georgia and Juan de Fuca and the Puget Sound.**

turn, we get a better sense of the reality that human culture is but a chapter in earth history.

The book is organized in three parts. Part I is background geology, consisting of chapters explaining large-scale earth structures and rock types. In Part II, the various geologic formations in the Islands are described, and interpreted in terms of plate tectonics. Also presented is the story of continental glaciation of the Islands. Part III of the book is a field guide, organized by island to island, emphasizing interesting places accessed on public lands. Also included in this part is a map of all photo localities.

The geology presented here is based on published materials stemming from many decades of research, mainly by faculty and students of the University of Washington and Western Washington University. The first comprehensive study of the Islands was the 1927 University of Washington PhD thesis of Roy McClellan. Geologic references are listed at the end of the book.

As a member of the Geology Department at Western Washington University since 1966, emeritus since 1999, the San Juan Islands have been at my doorstep for a long time, beckoning geological and recreational pursuits. In earlier years, my studies were mostly in the western Cascades, mapping and analyzing the giant thrust sheets that underlie this area. Similar work was going on in the San Juan Islands by numerous faculty and students at the University of Washington. At a certain point in the late 1980s, as rock types, ages, and structure became increasingly known, we all realized that the same assemblage of rocks occurs in both the San Juan Islands and the western Cascades. Then, of course, more progress could be made by working both regions together—which drew me into the Islands.

A breakthrough technology for unravelling the history of the San Juan Islands came my way in 2002 when my wife and I began wintering in Tucson. George Gehrels at the University of Arizona, finding that I was in the neighborhood, invited me to work in his dating lab. He was, and still is, at the cutting edge of age analysis of rocks using the trace mineral zircon, common in igneous rocks and sandstones. In my first tour of the lab, he took me into the mass spectrometer room where an undergraduate student sat at a computer, found a zircon grain in the microscope view, pressed a button, and within a minute had the age in millions of years pop up on the screen. I thought, I can do this! And over the years since then, we have greatly increased the age data-base for the San Juan Islands, in turn addressing many aspects of the geologic evolution.

Understanding of the complex geology of the San Juan Islands is in a constant state of upgrade. This book is a marker in that process and covers the main findings up to today, but tomorrow will bring new insights.

# ACKNOWLEDGEMENTS

Over the decades that I have worked on San Juan Islands' geology, my associations with many colleagues of like interest have been of fundamental value. In particular, Liz Schermer and Clark Blake, on trips to the Islands with me, have helped me understand the rocks and been a sounding board for models of origin. However, notably, when it comes to difficult questions of where disparate parts of the San Juan Islands originated and how they came together, we each have our own ideas (the "big picture" remains uncertain).

In developing this manuscript, numerous people have come forward with assistance. My brother Peter travelled out from Minnesota for a week to share rental costs of a cabin cruiser, and to drive the big boat around the Islands, gaining me access to photogenic rocks. He also, a writer by profession, gave me much help in getting started on the manuscript.

The style of presentation and accuracy of the science have been upgraded by a number of geologists. Working over the manuscript cover-to-cover were Jim Talbot, Clark Blake, and Michael Yeaman. Consulting on chapters related to their expertise were Doug Clark (glaciation), George Mustoe (Chuckanut Formation), Sue DeBari (Fidalgo ophiolite), and Thor Hansen (paleontology).

Because a primary purpose of the book is to bring geology to non-geologists, I enlisted my wife, Linda, and sister-in-law, Ellen Brown, to help me tone down the technical language and develop a more readable text – both of whom poured many days into this project.

Brendan Clark of Village Books designed the book cover and guided me in formatting the manuscript.

*Fossil Bay, Sucia Island*

# PART I – BACKGROUND GEOLOGY

## CHAPTER 1

# *TECTONIC CONCEPTS*

If you have spent much time on the Pacific coast of either of the Americas, Asia, or New Zealand, you have had an earthquake experience; you have truly felt the earth move (eerie if you are in the countryside, scary if in town). These earth movements, on a global scale and through geologic time, are the subject of "tectonics". Rocks and structures of the San Juan Islands preserve a fascinating tectonic record of tens of miles of vertical movements and thousands of miles of horizontal translations.

From fossils, rock types, rock ages, and preserved ancient magnetic signatures we deduce that different parts of the San Juan Islands were made in Asia, Scandinavia, and (maybe) Mexico. At first thought such large translations of rock terranes seem impossible because faults along which rocks are displaced move on average only a few inches a year. But we have lots of time to play with—hundreds of millions of years. And, since the 1960s we have a global tectonic model for displacements that explains so much about how rocks move on a grand scale. This is the field of "Plate Tectonics", the heart of this chapter.

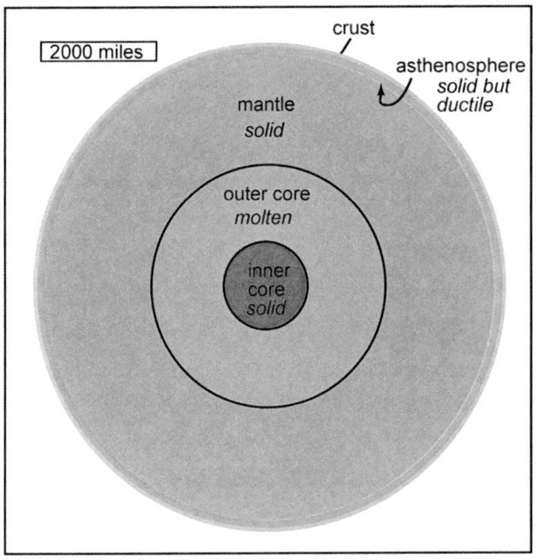

The deep structure of the earth is obviously far beyond reach of direct human observation. The distance to the earth's center is about 4000 miles, whereas the deepest drill holes go down only about 6 miles. Mainly from earthquake wave velocities, which delineate breaks in rock type, we divide the earth into the following zones: crust ~3-50 miles thick, mantle 1900 miles thick, and core 4000 miles across (Figs. 3, 4). We mostly can see what the crust is made of from drill holes, deep mines, and areas of crustal uplift, which can amount to as much as 100 miles.

**Fig. 3 Diagram of the deep structure of the earth. This model is based mainly on the measured velocity of earthquake waves that travel through the deep earth. See Fig. 4 for details of the crust and upper mantle.**

In the oceans the crust is thin, 3-6 miles, and made of basalt with a thin sedimentary cover (Fig. 4). The continental crust is thicker, generally 20-25 miles, and is composed of granitic, volcanic, and sedimentary rocks, and metamorphic equivalents of these altered by heat and pressure. The base of the crust is marked by a jump in earthquake wave velocity corresponding to a change in rock type passing into the underlying mantle. This boundary is called the "Moho", named after the Croatian seismologist Andrija Mohorovičić who discovered it in 1909.

Rock in the outer mantle is denser than crustal rock, is composed mostly of the iron-magnesium rich silicate minerals olivine and pyroxene, and is termed "peridotite". Deeper in the mantle other Fe-Mg minerals prevail. We see peridotite on the earth's surface owing to huge uplifts (an example is the Twin Sisters Range in the North Cascades), and the mineralogy occurs in a variety of stony meteorites so we know it has a cosmic abundance.

The earth's core is thought to consist mainly of iron and nickel based on seismic wave velocities, density calculations and the known abundance of these metals as components of meteorites. Earthquake wave velocities tell us that the inner part of Earth's core is solid and the outer part molten.

Elsewhere throughout the body of our planet there are not large zones of molten rock. The igneous rock we see in volcanoes or in subterranean bodies comes from relatively small local zones of melting in the upper mantle.

Critical to our understanding of crustal tectonics is the finding of a depth zone in the upper mantle of relatively low velocity transmission of seismic waves. This "low velocity zone" is interpreted to constitute a weak, easily sheared, close to melting but still solid, part of the mantle; it is termed the "asthenosphere" (Figs. 3, 4). Above the asthenosphere is a relatively rigid zone termed the "lithosphere", which constitutes the uppermost part of the mantle and the overlying earth's crust. The asthenosphere is mostly solid, but viscous, and over a long period of time can flow due to being near its melting temperature. This property is very significant for understanding up and down movements of the earth's surface geology.

Rock loads piled on the crust, such as volcanoes or thrust sheets, or buildup of continental ice sheets, depress the lithosphere into the asthenosphere. As these materials are removed by erosion, faulting, or melting, the earth's crust rebounds—a type of "floating equilibrium" exists between the lithosphere and asthenosphere. Asthenosphere weakness also allows a decoupling with the lithosphere during lateral shear, thus rigid plates of the lithosphere are able to slide as much as thousands of miles on this slippery substrate.

The modern Pacific Ocean basin and its margins tell us much about global tectonic processes (Fig. 5). Active geology around the Pacific Rim is well known, especially by locals to the area, to involve live volcanoes and seismic zones—the "Ring of Fire". Coupled with these features are deep sea trenches. A large-scale tectonic model for these ocean margin features got a big start with the discovery by American seismologist Hugo Benioff in the 1950s, that earthquakes of the Pacific margin lie in a plane dipping away from the Pacific and under the continents (Figs. 4, 5). As it was understood that earthquakes are related to fault motion, then it was apparent that in these "Benioff zones" very large-scale faults separate the Pacific Ocean from the continental land masses.

Another piece of the tectonic model came into place in the 1960s with the understanding from fossils in sea floor sediments and patterns of magnetic anomalies in oceanic basalt that the age of ocean crust increases systematically away from ocean ridges (Fig. 5), and therefore must be created at the ridges. Subsequent deep ocean studies by manned submersibles and remote sensors document active sea floor volcanism along the ocean ridges. (Spectacular video clips are seen on TV nature shows of the hydrothermal vents and a remarkable ecosystem dependent entirely on chemical energy emitted along ridges.)

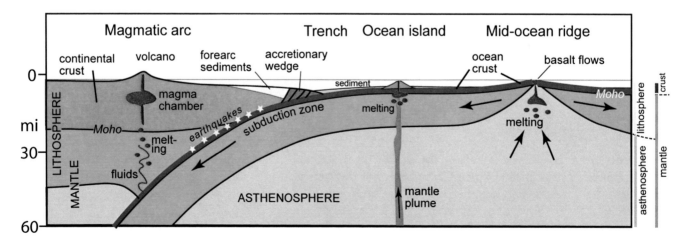

**Fig. 4 Schematic illustration of plate tectonic features across an ocean basin and convergent continental margin.**

These findings led to the "plate tectonic" revolution of geologic theory. In this explanation of global scale earth movements, the ocean crust is generated at the ridge, it moves off the ridge travelling somewhat down slope as a coherent plate pulled by gravity along the top of the asthenosphere, and slides toward the ocean margin where it dives into the deep earth along what is now termed a "subduction zone". The downward motion at the ocean margin creates a trench, which may or may not be filled with continent-derived sediment.

In the subduction zone the increased heat and pressure on subducted rocks releases water and other volatiles that rise into the overlying mantle and cause melting of the hot peridotite. The molten rock (magma) rises and is emplaced high in the over-riding plate as volcanoes at the surface, and magma chambers at depth which crystallize into plutons. These igneous rocks, volcanoes and plutons, typically lie in arcuate belts, a geometry resulting from the intersection of the ocean plate and earth curvature, and are termed "arcs". Some arcs are developed within the margin of the ocean basin, and are called "island arcs"; others form within the continental margin, and are termed "continental arcs". For example, the active magmatism in the Andes of South America comprises a continental arc, whereas arcs in the western Pacific, such as the Mariana Islands, are island arcs (Fig. 5).

Additional ocean features important to the story of continental growth are oceanic islands and seamounts (Figs. 4, 5) that occur within the ocean plate and are not related to subduction, such as the Hawaiian chain. These ocean islands are volcanic, fed by a source of basalt magma generated in the upper mantle by melting of an apparently deeply rooted mantle plume, termed a "hot spot". The hot spot stays fixed in position while the volcano it generates eventually rides off the source as part of the

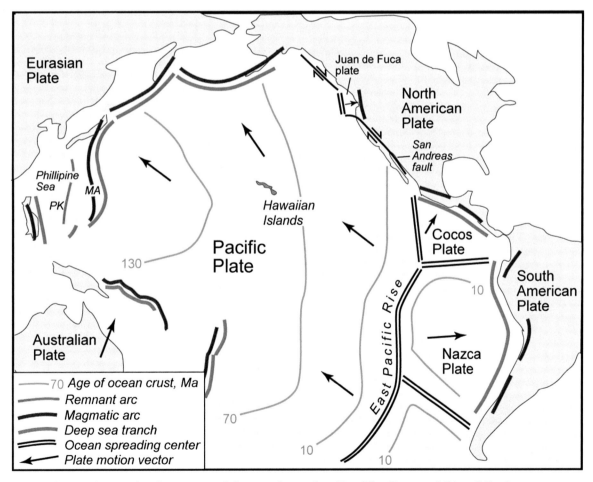

**Fig. 5 Tectonic elements of the modern day Pacific Ocean. MA = Marianas Arc, PK = Palau - Kyushu remnant arc, Ma = million years old.**

moving oceanic plate. Thus in the Hawaiian Island chain the active volcano is on the Big Island at the south end of the chain, and the rest of the islands are extinct volcanoes that are progressively older going northwest from island to island as the Pacific plate slides northwest.

Continental growth can occur in the zone of convergence between the oceanic and continental plates, in what is known as the "accretionary wedge" (Fig. 4). Here, while much of the ocean plate is subducted deep into the mantle, some parts are "under-plated" (peeled off and stuck) on the continental margin as a result of the subduction zone stepping back. These accreted materials can be ocean crust, ocean island volcanoes, island arcs, even micro-continents that have been broken away from their place of origin.

As easily imagined, the accretionary wedge records a great amount of rock strain in the form of faults. Also commonly formed is a penetrative planar fabric, "cleavage", due to shear and flattening associated with ductile deformation. The rocks can look smeared. In the subduction zone, rocks carried to some depth before accretion and uplift are subjected to high confining pressures in addition to shear stress. Confining pressure is the pressure that comes from all sides, as in the deep ocean or in the earth where solid rocks can flow over time. The rocks also are considerably heated due to the ambient high temperatures at depth in the earth. The original minerals in these rocks recrystallize to other minerals that are stable at the elevated temperature and pressure.

In some uplifted subduction zones, the depth and thus pressure was even great enough to form diamonds from organic materials carried down from the surface. Burial of these rocks to a depth greater than 75 miles is indicated from experimental studies of the stability of diamond vs. its low pressure form graphite. Thus, we find that accreted rocks are typically changed in texture and mineralogy due to strain and elevated temperature and pressure. These are metamorphic rocks. For the purpose of understanding how accretion works, the metamorphic rocks provide a good tool for evaluating direction of transport and amount of burial during metamorphism.

Understanding of the plate tectonics of the Pacific basin (Fig. 5) gave us a breakthrough insight into the origins of bedrock along coastal western North America, including the San Juan Islands. As explained above, ocean crust is generated at ridge systems and moves away toward the ocean margins in both directions, approximately equally. In the modern Pacific the ridge system is the "East Pacific Rise". The Pacific Plate moves northwest and plates on the east side, Nazca, Cocos and Juan de Fuca, move to the east and northeast. Plate velocity is about 3 inches per year. The Pacific Plate is huge and dates back in its oldest part to about 170 million years. There is no such old component of the eastern plate. The Nazca and Cocos plates are small and relatively young. Along the coast of North America the only eastern plate is the very small and young Juan de Fuca Plate.

What happened to all the plate material that was generated by the East Pacific Rise ridge system, and why is the ridge system cut off in the southwest North America? Following plate tectonic logic, we interpret that most of the eastern plate generated on the East Pacific Rise ridge system was subducted under North and South America. In fact convergence of the subducting plate against North America was so robust that in California the ridge itself was carried under the North American plate. In the absence of the eastern plate, the boundary now between the North American and Pacific plates is the San Andreas fault, which slides coastwise moving westernmost California north.

The modern Pacific plate tectonic configuration tells us that ocean crust once as big as the entire breadth of the Pacific Ocean, 4000 miles wide, has been subducted under western Washington, and is still happening; the Juan de Fuca plate is a small remnant of the eastern plate which has mostly disappeared. We have named this ancestral ocean plate the "Farallon Plate". The entire west coast of North America is a repository of accreted chunks of oceanic stuff scraped off the Farallon Plate: ocean crust, island arcs, ocean island volcanoes, and even bits of foreign continents. This is the bedrock of the San Juan Islands.

# CHAPTER 2
# *GEOLOGIC TIME AND DATING*

In modern times we understand the earth to have had a profoundly long history, and our own presence on earth to be a tiny fraction of this history. Earth was formed from cosmic debris some 4500 million years ago, and evolution produced humans from ancestral primates only about 150,000 - 200,000 years ago. If we strung out a line along the length of a football field to represent the duration of earth history, the measured length of humans on earth would be the thickness of about two human hairs.

Understanding of the age of the earth, and unraveling earth history from rocks has been a long process in human endeavor. Imagine being a Middle Ages outdoors person exploring around in the Alps and coming upon a clamshell embedded in rock. Following current thought of the time, you would know that the mountains are eternal and that the shell was created by a mountain animal living in the rock. Nowadays we would marvel at the evidence for seafloor sediment containing shellfish, hardened into rock, and uplifted to form a mountain. In the Age of Enlightenment in the 16th and 17th centuries, principles of the scientific method were embraced; it was noted that shells embedded in rock looked like seashells and therefore the rock itself came from the sea. A Danish catholic priest, Nicolas Steno, led the way during this period in establishing the principles of stratigraphy: Layered rocks are commonly sedimentary in origin, the layers are progressively younger going upward in the sequence, the layers are broadly horizontal and represent time markers, and fossils in the rocks help define layers of similar age.

The role of fossils in delineating earth history (Fig. 6) was greatly developed in the 17th and 18th centuries, and in the early 19th century the term "paleontology" came to be used to describe the scientific study of fossils. Recognition that populations of ancient organisms changed with time due to extinctions and evolution gave them special value in establishing the relative age of rocks. On this basis development began of a "stratigraphic chart" delineating global periods that are marked by distinctive changes in fossil populations. Names are given to the various time periods based on a type locality where the definitive fossil assemblage was first found. For example, the Devonian period is named for fossiliferous strata in Devon, England where distinction of this period was established in 1840. A modern version of this chart is given in Fig. 7. Refinements occupy paleontologists today.

**Fig. 6 Ammonite fossil from sandstone on Barnes Island. This genus, Reesidities, lived during a relatively narrow time range within the Cretaceous Era, calibrated to an interval of about 89-91 Ma (from Haggart et. al 2001).**

Aside from the progress made in establishing the *relative ages* of rocks, the *absolute age* was a different problem. Nineteenth century scientists did the best they could with tools and concepts they had. How much time is involved in creating a mountain made of thick sequences of ocean-deposited strata followed by miles of uplift? And how long did it take for evolution to create the life forms preserved as fossils in a stratigraphic sequence? As far as could be understood, long slow processes were involved. An earth age of at least tens of millions of years was recognized.

The discovery of radioactive elements by the Polish physicist Marie Currie in 1898 opened the door to precise absolute age dating of rocks. Certain elements and isotopes of elements are naturally unstable; the atoms give off radiant energy and break up into other elements at a very constant rate. Of course this science led to development of nuclear power plants and bombs, both problematic. But geologists adapted the process as a means of age-dating rocks. The "geochronologist" measures the

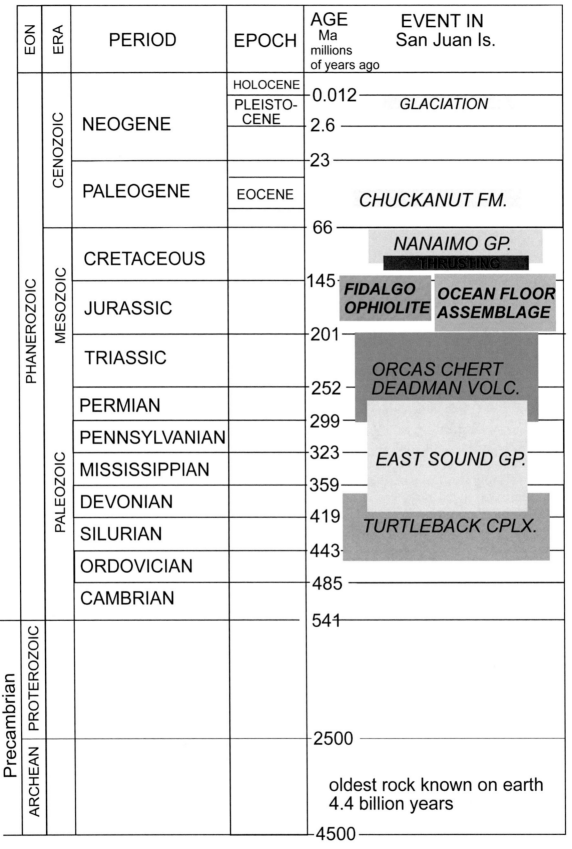

Fig. 7  Stratigraphic chart, with age ranges of rock units of the San Juan Islands. This diagram is based on the "International Chronostratigraphic Chart" of the International Commission on Stratigraphy 2013.

amounts of the unstable element present in a rock and the new element produced, and knowing the decay rate, calculates the age.

By the 1920s measurement of radiogenic ages in earth rocks pushed the known earth age out to several billion years. Today, the oldest known earth rock, in Australia, yields a 4.4 billion year age. An age of the earth of 4.5 billion years is recognized from ages of meteorites, which are thought to be part of the material out of which earth was formed. Procedures for making these measurements are the subject of physics and chemistry; in the modern day amazingly precise ages are obtained in sophisticated laboratories. A convenient shorthand for giving the age of a rock in millions of years before present is simply Ma (or Ga or Ka if the age is in billions or thousands of years).

For the San Juan Islands, three chemical systems that have greatly enriched our understanding of the geologic history are: 1. uranium decay to lead, 2. breakdown of an unstable isotope of potassium ($^{40}$K) to argon gas ($^{40}$Ar) and 3. decay of the unstable isotope of carbon ($^{14}$C) to the stable isotope of nitrogen ($^{14}$N). For uranium-lead geochronology, zircon is the favored mineral as it takes up a small amount of uranium when it crystallizes from magma, and strongly holds the daughter product, lead, in the crystal lattice for measurement in the laboratory (Fig. 8). The decay is slow, so this geochronometer is accurate for rocks older than about 10 million years. Zircon is particularly useful because it is widespread in trace amounts in igneous rocks. Zircon is also of great value in sedimentary rocks where it occurs as sand grains that yield the age of the provenance from which the sand was eroded, and it also limits the maximum possible age of the rocks.

The modern technique of age-dating zircons uses a method of spot analysis based on zapping a small part of a crystal with a laser beam, followed by analysis of the vaporized zircon in a mass spectrometer which measures the various isotopes of interest. As complicated as the machinery and science are, the age collecting is pretty simple: find a

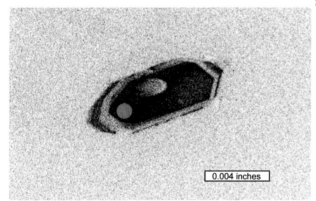

**Fig. 8 Zircon crystal extracted from igneous rock of the Turtleback complex in Deer Harbor, Orcas Island. Uranium/lead isotope analysis indicates an age of 380 million years. The photograph is an electron microscope image showing chemical zoning. The red spot marks a portion of the crystal vaporized by a laser beam. The zircon vapor passed into a mass spectrometer for isotope analysis, at the University of Arizona LaserChron lab.**

spot on the zircon crystal, press a button to activate the laser, and shortly the age appears on the computer screen. For a San Juan Island field geologist (that would be me) who has exhausted all other tools to get an age, sitting at this computer and pressing the laser button is high drama as so much earth history unfolds.

Potassium to argon decay, a good geochronometer for materials 100,000 years and older, has been useful in the San Juan Islands for obtaining ages of mica in metamorphic rocks.

Carbon 14 exists in the atmosphere in a constant proportion with the stable isotope carbon 12. This proportion is carried into organic material at growth. As time progresses, the $^{14}$C decays to $^{14}$N, and thus an age can be determined from measuring the two carbon isotopes and the stable nitrogen isotope, and applying the known decay rate. This technique is good for ages in the range of about 100 to 60,000 years. For the San Juan Islands the $^{14}$C dating of wood in glacial deposits has been the key tool in defining the age of burial by the continental ice sheet, and its departure.

The stratigraphic chart of Fig.7 is laid out based on fossil distinctions of time units. Radiometric age-dating adds an absolute chronology to the chart, and is fine-tuned as more data become available.

# CHAPTER 3
# ROCK TERMINOLOGY AND ORIGINS

Every rock you see has a story to tell, and often the history we can deduce from an outcrop or even single specimen is mind-bending. In this section we'll cover some terminology and theory as a foundation for discussion of rock origins in the San Juan Islands. Considerable interest in understanding the rocks pertains to their plate tectonic history, and therefore the presentation here is tilted toward that end. This is a much abbreviated view of mineralogy and petrology, and the reader may well wish to expand the information by reference to textbooks or by simply checking out certain topics on the internet.

Igneous rocks

Of the three major classes of rock, igneous, sedimentary, and metamorphic, the igneous rocks get by far the most press. We see in the news, or have in our backyard, active volcanoes blasting out red hot blobs and streams of lava. The definition of igneous rock is that it has a molten origin; it comes from solidification of magma. The magma itself is generated in the outer part of the earth, both in the upper mantle and in the lower parts of the crust. By far, most of the outer part of the earth is solid, so melting is a special phenomenon.

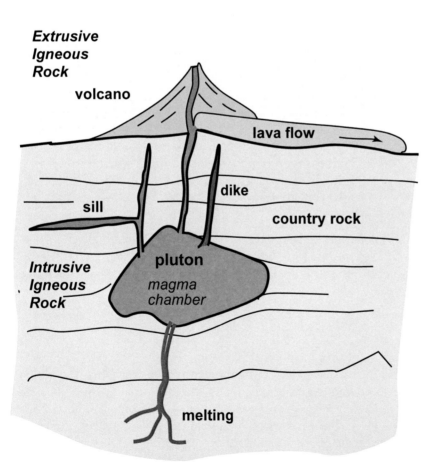

**Fig. 9 Diagrammatic sketch of intrusive and extrusive settings of igneous rock formations.**

Melting in the mantle occurs at varying depths beneath ocean ridges, arcs and ocean islands, giving rise to basalt magma (Figs. 4, 9). The mantle rock that melts is peridotite, consisting mainly of the minerals olivine and pyroxene. It's a partial melting process that takes the pyroxene and leaves behind olivine. Some magma generated in the upper mantle erupts on the earth's surface as basalt or andesite. In other places the magma ascends into the crust, but stalls, cools and crystallizes below the ground surface, either as sheet-like bodies termed sills and dikes, or in the form of large magma chambers—some thousands of cubic miles in volume. A modern day example is the vast body of magma under Yellowstone Park detected by seismic studies.

Magma that crystallizes more slowly, usually as an intrusion at depth, tends to be coarse-grained—that is the crystals can be easily seen. Magma that cools fast, at shallow earth depths or on the surface, may be either glass (obsidian) or fine-grained crystalline rock. Basalt and the coarse-grained equivalent, gabbro, are dark colored and relatively rich in iron and magnesium and low in silica compared to the spectrum of igneous rocks (Figs. 10-12). The more silica-rich rocks, such as granite which is light colored, can form by either melting of deep crustal rocks at the base of continents, or by fractionation of original basalt magma.

The fractionation process is probably most important in producing the array of light and dark colored igneous rocks seen in the San Juan Islands (Figs. 11, 12). These rocks originated in a magma chamber at depth in the roots of an arc where crystal settling caused original andesite or basalt magma to be split into parts of differing mineralogy. As magma cools, it crystallizes over a temperature range, and different minerals crystallize at different temperatures. Early-formed crystals are pyroxene, olivine, and calcium-rich feldspar. These can sink in the magma and settle on the floor of the chamber as a kind of igneous sediment, forming a layered gabbro (Fig. 13). The remaining melt is depleted in iron and magnesium and enriched in silica and crystallizes relatively abundant amounts of feldspar and quartz, such as in granite. All degrees of this process occur, producing a

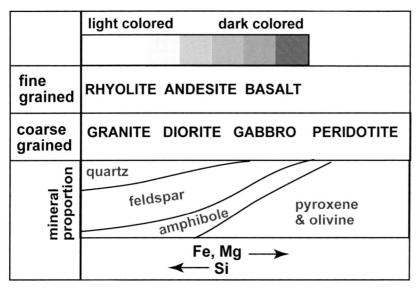

Fig. 10 Igneous rock classification chart. The fine-grained rocks, termed volcanic, crystallized quickly from magma at the earth's surface or at shallow depth. The coarser rocks crystallized slowly at depth, and are termed plutonic.

Fig. 11 Granite surrounded by gabbro in the Turtleback Complex on Ship Peak, Orcas Island. In the granite, feldspar is pinkish and quartz slightly grey. The gabbro is composed of specks of light-colored feldspar and an abundance of dark pyroxene, or amphibole.

Fig. 12 Diorite from the Turtleback Complex at Deer Harbor on Orcas Island. This rock is composed of dark colored amphibole and light colored feldspar.

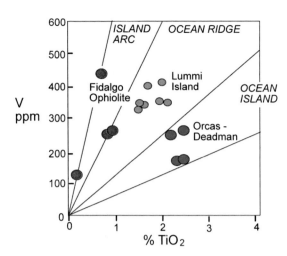

**Fig. 14** Plot of vanadium vs. titanium oxide as measured by chemical analyses of basalts in the San Juan Islands. Tectonic settings are distinguished.

**Fig. 13**, *left*, Sedimentary layering in gabbro caused by differential settling of light and dark minerals in a magma chamber. The layers were originally horizontal and have been tipped up on end. Fidalgo ophiolite, Alexander Beach area, Fidalgo Island.

continuum of rocks ranging across the entire chemical spectrum of Si-rich to Fe- and Mg-rich (Fig. 10).

Some fractionated magma crystallizes in the magma chamber, but much erupts as volcanic rock. The volcanic rock occurs as layers of lava and also as fragmental explosive deposits of coarse breccia and fine-grained tuff, as observed in the 1980 catastrophic eruption of Mt. St Helens. Much igneous rock occurs in the San Juan Islands as dikes. These mark feeders to higher level rocks, including volcanoes, where magma has squeezed into and travelled through cracks. The dikes typically occur as injection complexes of several phases in the same outcrop (Fig. 11), marked by light and dark colors reflecting their variance in composition. Also seen in the Islands is the base of the magma chamber itself where gabbro exhibits a layered structure formed by sedimentation of early-formed crystals settling onto the floor of the chamber (Fig. 13, see also Chapters 5 and 9).

Basalt magma generated in the mantle erupts in three tectonically diverse oceanic settings, as mentioned above: ocean ridge, ocean island, and island arc. What evidence do the rocks hold that allows us to identify these settings?

Arc rocks are the easiest to deduce. The dominant igneous rock found is what we term "intermediate", that is between the end members of basalt and rhyolite. This would be andesite or diorite depending on grain size. By far the dominant igneous rock of ocean ridges and ocean islands is basalt, not obviously different from one setting to the other. However, chemical analysis shows that the ocean island basalt is consistently enriched in titanium. Chemical plots of San Juan Islands basalts provide discrimination for all three tectonic settings (Fig. 14).

Metamorphic Rocks
Metamorphic rocks are a treasure in that they bring to the earth's surface clues as to what has happened far below where humans can go. Sedimentary and igneous rocks buried deep in the earth undergo a metamorphosis, not of supernatural origin, but caused by heat and pressure that modifies the mineralogy and texture. Mineralogic change in these metamorphic rocks represents chemical reactions forced by elevated temperature and pressure conditions due to burial and/or proximity to a big heat source such as a pluton. Experimental geologists learn about metamorphic processes by

**Fig. 15** Limestone from the old McGraw - Kittinger quarry on Orcas Island. The rock is a marble, fully recrystallized from the original mineral calcite, to the high pressure metamorphic polymorph aragonite; both minerals are forms of $CaCO_3$. Original textures of the calcite limestone appear as gray material crossed by light veins and dark carbonaceous fracture fillings. The aragonite is evidenced by the thin reflective bands crossing the whole sample. These are fine crystallographic lamellae (twins) showing up on a broad cleavage surface in the aragonite structure.

This rock has had a remarkable history. It originated on the sea floor by accumulation of shellfish, was then reconstituted during sedimentary compaction obliterating most of the shell structure, next was subducted to as much as 12 miles or more below the earth surface as the limestone terrane collided with the continental margin, and finally was uplifted and eroded to be exposed at the earth's surface.

placing capsules of powdered rock into super-strong steel containers, "bombs", placed in an oven. Water pressure is pumped up in the bombs to many thousands of times atmosphere pressure, and temperature is raised to a thousand degrees Fahrenheit or more. After days or months, when the experiment is over and the capsule is opened, voila! new minerals are formed, metamorphism has happened.

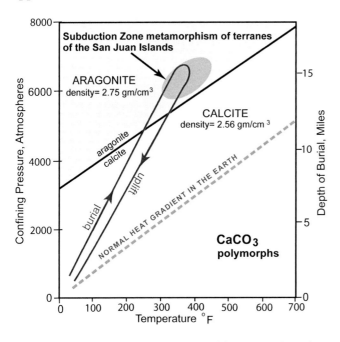

**Fig. 16** Pressure - temperature diagram showing the stability of aragonite relative to calcite, deduced from laboratory experiments.

Green color in igneous rocks is a common metamorphic effect, representing alteration of igneous Fe-Mg minerals such as pyroxene, olivine, and hornblende to metamorphic chlorite, epidote, serpentine, prehnite, or pumpellyite. These five metamorphic minerals are various shades of green, not easily distinguished in fine-grained metamorphosed volcanic rocks that we simply refer to as "greenstone".

A very cryptic, but tectonically exciting metamorphic mineral change is that of calcite to aragonite; both are forms (polymorphs) of $CaCO_3$ (Fig. 15). The aragonite is the high pressure, denser, form (Fig. 16). An analogy is the relation of diamond to graphite, in which at high pressure the denser polymorph, diamond, is stable. Aragonite looks pretty much like calcite in being of the same color, having a good cleavage surface, being scratchable with a knife, and fizzing in dilute

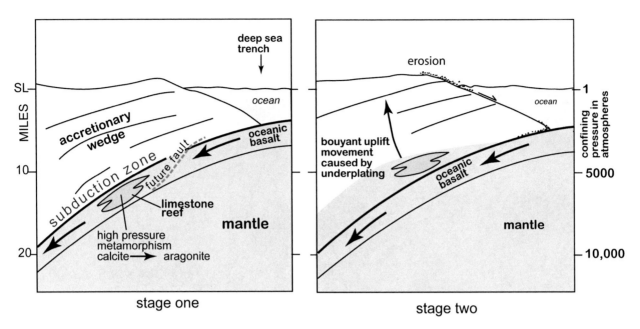

**Fig. 17** Schematic section showing plate motion of ocean crust into the subduction zone and subsequent transfer to the accretionary wedge. The subducted slab breaks along a fault and steps down, leaving a cut-off portion of the slab above, thickening the wedge. The cut-off part of the slab rises through the wedge as more material is added to the bottom and erosion denudes the top. Calcite conversion to aragonite occurs in the subduction zone below about 12 miles depth, caused by high confining pressure. The buoyant uplift of aragonite in the wedge is fast enough that conversion back to calcite does not happen.

hydrochloric acid—all geology 101 tests. However, the atomic structure is much different and can be identified by x-ray diffraction analysis, as was done by Joe Vance of the University of Washington in discovery of this mineral in the San Juan Islands. From Vance's and later studies we know that aragonite occurs widely in the older rocks of the region, noted specifically in later chapters.

The tectonic interest in aragonite comes out of the pressure - temperature conditions where the mineral is stable (Fig 16). From associated metamorphic minerals, the prehnite and pumpellyite mentioned above plus others, we surmise that the temperature was in the range of about 300 - 450$^0$F. For aragonite to have formed at this temperature, the pressure was greater than about 6000 times our atmospheric pressure. We relate metamorphic pressure to depth of burial in the same sense as water pressure relates to depth below the surface. Rocks at depth are warm, plastic and slowly flow, responding to depth by a squeezing in from all directions. We call this *confining pressure*, as opposed to *shear pressure*. Confining pressure increases with depth of burial such that at 6000 atmospheres, burial was 12 miles. These figures give us a geothermal gradient of about 30$^0$F per mile going to depth in the earth. We know from geothermal studies that this is a very odd gradient—a normal earth gradient is about 50-60$^0$F per mile, as for example measured in deep mines.

How can we understand why the San Juan Islands rocks became so deeply buried, and why this burial was so unusually cold for earth conditions? Plate tectonics comes to the rescue (Fig. 17). Subduction zones carry surficial rocks down to great depths rather quickly (inches per year). Confining pressure on these subducted rocks rises almost immediately but heat flow into the rocks is slow, thus high pressure (P) - low temperature (T) conditions prevail. There is more to this however. We can see how an initial P-T condition could allow stability of aragonite, but eventually the rocks would heat up to a normal temperature for that depth and bring them into P-T conditions of calcite stability, which we know from experiments would readily form. So, an added feature to the tectonic

Fig. 18 A bedded sequence of light colored sandstone and darker siltstone is intensely broken and sheared by ductile deformation. Irregular white quartz veins cross the deformation fabric. Beach below Cattle Point parking area, San Juan Island.

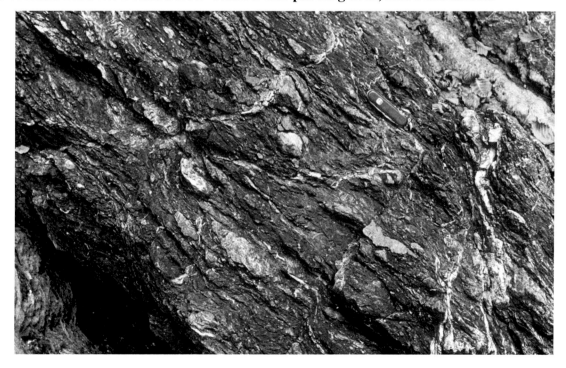

Fig. 19 Highly broken formation in a fault zone, showing a scaly planar shear fabric, cross-cut by later quartz veins. Near Rosario resort on Orcas island.

story is that the aragonite rocks were uplifted quickly—virtually shot out of the subduction zone (again, a few inches a year). This is the metamorphic history of much of the San Juan Islands. (By the way, diamonds in our possession also are out of their stability range, wanting to be graphite. Like aragonite, they came up quickly, blasted up from the earth's mantle in a volcanic eruption.)

Besides mineral changes inflicted on (enjoyed by?) rocks during deep burial due to temperature and high confining pressure, shear caused by plate motions changed the shape of rocks and imprinted fabric related to ductile flow, flattening, and brittle fracture. Silly putty gives us a nice analog for rock deformation—fast deformation is brittle and slow deformation is ductile. Heat also facilitates flow over fracture. Some rocks in the San Juan Islands show strongly disrupted bedding crossed by countless small ductile shears (Fig. 18). Near major fault boundaries between thrust sheets, the rocks are crushed and mixed on brittle fractures (Fig. 19). Effects of deformation are seen throughout rocks of the Lummi Formation where penetrative ductile shear has created a planar fabric, termed foliation, defined by small shears and parallel orientation of platy minerals. Besides shears, folds are also a common expression of compression in metamorphic rocks.

### Sedimentary rocks

Sedimentary rocks are born of the local geology that supplies sediment, and are strongly influenced in texture and composition by the environment in which they are deposited. Thus we find good information in sedimentary rocks about the ancient landscape and the tectonic setting of their formation. In many ways, study of sedimentary rocks to discern the past geological environment is

| CLASTIC ROCKS (fragmental) | | |
|---|---|---|
| | **grain size** | **clasts** |
| Mudstone | <1/256 mm | mud |
| Siltstone | 1/256 - 1/16 mm | silt |
| Sandstone | 1/16 - 2 mm | sand |
| Conglomerate | > 2mm | pebbles, cobbles, boulders |

Greywacke = sandstone with components of mud, silt and a mixture of mineral grains and dark rock fragments

Quartz sandstone = sandstone composed of more than 90% quartz grains

Arkose = sandstone with more than 30% feldspar grains.

Breccia = coarse clastic rock with grains > 2mm and very angular, not rounded as in conglomerate; forms at the base of cliffs

Turbidites = interbedded greywacke sandstone and siltstone, commonly graded upward from coarser to finer grained, inferred to be deposited as a turbid slurry

| ORGANIC SEDIMENTARY ROCKS | |
|---|---|
| Limestone | accumulation of calcium carbonate shells, $CaCO_3$, as oceanic reef structure, or bedded debris, typically associated with ocean arc volcanoes. |
| Chert | siliceous radiolarian ooze on the deep sea floor reconstituted as dense hard quartzose beds, associated with ocean ridge basalts |
| Coal | in the Chuckanut Formation, derived from compacted swamp plant material, associated with river deposits |

**Fig. 20 Classification chart of sedimentary rocks found in the San Juan Islands.**

comparable to archeological study of ancient middens (garbage dumps) as a window into pre-historic human society.

Sedimentary rocks are formed by hardening, termed lithification, of accumulations on the earth's surface of debris formed by a variety of means: weathering and erosion of other rocks, organic growth, and chemical precipitation. This debris is all around us as mud, sand, gravel, decayed plant material, shell deposits, and salt precipitates in briny seas. The lithification process involves compaction and squeezing together of debris particles and expulsion of water. Sand and gravel deposits additionally are hardened by chemical precipitation of mineral cement in pore spaces. Shell accumulations, after deposition and burial, commonly recrystallize into a denser aggregate. All aspects of lithification happen close to the earth's surface at near surface pressure and temperature. Deeper in the earth, as sediments are buried, high temperature and pressure cause further change, but this effect is termed metamorphic.

For a start, we look at the sedimentary rocks to discern whether their origin was fragmental (clastic) or organic (Fig. 20). For the clastic rocks important clues about the origin come from the grain size, roundness, and mineralogy. All these features have meaning for the environmental and tectonic conditions of formation of the rocks.

Greywackes (Fig. 21), so abundant in the San Juan Islands, are a kind of sandstone composed of chemically "immature" mineral grains and rock fragments, indicating short exposure to the atmosphere that given time will dissolve (chemically weather) almost all minerals except quartz. A limited experience of these grains in being rolled and bounced by transporting water is also indicated by low degree of rounding. Greywacke sands did not languish in their travel from outcrop to final resting place. From this we can interpret fast streams and relatively steep topography. From an abundance of volcanic fragments and chert grains in the San Juan Islands' greywackes, the landscape is deduced to be underlain largely by these rock types. The greywacke has mud and silt mixed in with the sand. This poor sorting of grain sizes indicates very little of the winnowing action one sees at

GREYWACKE

QUARTZ SANDSTONE

ARKOSE

**Fig. 21 Three sandstone types in the San Juan Islands with very different origins.** *Greywacke* **sand of the Ocean Floor Assemblage was eroded and transported quickly from a highland of oceanic chert and volcanic rocks. Triggered by earthquakes, turbidity currents carried the sediment from coastal areas out onto the deep ocean floor (Fig. 22).** *Quartz sandstone* **making up this rounded cobble from the Nanaimo Group originated by protracted weathering of bedrock that removed all other minerals in a low-relief tectonically stable continental setting.** *Arkose* **sand in the Chuckanut Formation is rich in feldspar and quartz and came from erosion of granite of continental origin. The sand was eroded from bedrock and transported relatively quickly, with limited exposure to chemical weathering in an area of some topographic relief.**

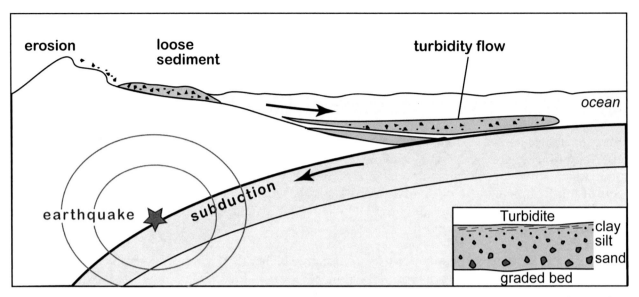

**Fig. 22 Probable tectonic setting and mechanism of turbidite formation. This scenario depicts a convergent plate margin where active uplift occurs above an accretionary wedge (as in Fig. 17) leading to rapid erosion of the arc as well as the accreted oceanic sedimentary and volcanic rocks. The eroded debris accumulates on the ocean margin as loose sediment until it is mobilized by earthquake activity and flows across the ocean basin as a dense, turbid current of water and debris of all sizes. As the turbid flow comes to rest, the coarser material settles first and the clay particles last. A size-graded sedimentary bed forms, typically of greywacke composition.**

beaches or in stream beds; thus the deposition of greywacke sediment is inferred to occur by settling from a slurry of unsorted materials. In a layered outcrop we commonly find a bed of sandstone grading upward to siltstone and mudstone, a bedding sequence interpreted to develop from differential settling rates. The settling rate of grains depends on their size—coarser settle more quickly. In this view, the slurry constitutes a dense, heavy turbid current that sweeps across the floor of the sedimentary basin where it eventually slows and stops, allowing the grains to drop out. We could surmise from this analysis the following history of greywacke turbidites of the San Juan Islands: The process begins with rapid erosion of a landmass composed of uplifted oceanic volcanic rocks and chert; then rock slides and stream transport carry the eroded debris quickly down the topography to accumulate as poorly sorted sediment amassed at the edge of a basin, as in a coastal river delta; and finally episodic shaking of the sediment loose from the delta, likely by an earthquake, sweeps the slurry of mud, silt and sand into deeper water to its final resting place on the floor of the basin. A stack of these graded beds in an outcrop represents multiple turbidity deposits and an environment repeatedly shaken by earthquakes (Fig. 22). The life history we interpret for greywackes is a perfect fit for an "active" margin, where plates collide.

The sedimentary antithesis of greywacke in terms of tectonics is pure quartz sandstone, which forms where topography is subdued, erosion is very slow, and chemical weathering has a long time to dissolve all minerals other than quartz. Purification of quartz sand probably requires more than one cycle of erosion, weathering, and sedimentation. The presence of quartz sandstone is a mark of a "passive" plate margin, where there is no plate boundary at lands edge. Such rocks occur in the San Juan Islands as exotic pebbles and cobbles (Fig. 21) in sedimentary bedrock of the Nanaimo Group, which otherwise consists of greywacke. We'll come back to this conundrum in Chapter 10.

The youngest clastic sedimentary rock in the San Juan Islands, found in the Chuckanut Formation, differs from the greywacke in interesting ways that help us understand a different tectonic setting. Sandstones are rich in feldspar and are termed arkose (Fig. 21). Grains were sorted by size;

winnowing happened. Instead of graded bedding features, we see cross-bedding indicative of deposition along sand bars in a stream. The abundance of feldspar indicates limited chemical weathering and a granitic source. Uplift and erosion of a broad granitic terrain implies a continental, not oceanic, setting.

Limestone is the graveyard of masses of shellfish and other marine animals. Most limestones in the San Juan Islands are closely associated with basalts and cherts, pointing to a far-from-land oceanic origin. This is in contrast to the vast limestone formations we find elsewhere in western North America deposited on the once submerged continent (for example in the Grand Canyon). In the San Juan Islands, limestones formed reefs associated with both ocean island and island arc volcanos, and in some places are interlayered with chert, suggesting a deeper ocean floor setting. The shellfish fossils provide critical information regarding rock age and where the sediment formed in the ancient geography of ocean basins and continents.

Chert is another common sedimentary rock in the San Juan Islands that contributes usefully to our understanding of the big picture of sedimentary environments and plate movement. In the San Juan Islands, chert occurs rhythmically interlayered with mudstone on a scale of each layer being an inch or so thick, and a bedded sequence being up to hundreds of feet thick. The chert layers are hard and resistant and stand out. The mudstone is soft and recessive and forms thinner layers. An outcrop surface of this rock is impressively banded, and it is referred to as "ribbon chert" (Fig. 23). In microscope view, and sometimes with a magnifying glass, one can see the small shellfish, radiolaria, that have accumulated to make this deposit. Radiolarian shells are siliceous; most have reconstituted during lithification to massive opal or quartz in the rock. The components of ribbon chert began in one part as airborne clay, and in another part as radiolarian plankton in near-surface ocean waters. The clay and plankton settled to the deep sea floor,

**Fig. 23. Ribbon chert at Rosario Head, Fidalgo Island. Hard, erosion-resistant beds of chert are separated by recessive layers of softer clay-rich sediment.**

constituting a clay and radiolarian ooze. The marked layering that we find now in these rocks apparently happened as a segregation process during compaction of the ooze, separating the siliceous material from the clay.

How does ribbon chert help us unravel San Juan Islands' geology? We know from modern day ocean sediment that the ooze accumulates slowly and in an environment where other sediment is not being deposited; thus the chert formed in a very quiet ocean setting, apparently far from eroding land. The radiolaria, because they evolved quickly, can be age-dated to a very fine timeline. From other parts of the earth, such as the Marin Headlands in California, radiolarian ages in ribbon cherts indicate a deposition rate of about three to five feet per million years. Thus, in measuring the thickness of a ribbon chert deposit in the San Juan Islands we can get an estimate of how long this quiet ocean floor setting was maintained. The age range, together with the generation rate of ocean crust at the ridge system, gives us an estimate of the breadth of ocean floor accreted in the San Juan Islands.

# PART II – GEOLOGIC HISTORY OF THE SAN JUAN ISLANDS

*CHAPTER 4*

# OVERVIEW

The geologic history of the San Juan Islands spans half a billion years. Most of the rocks came from a far distant homeland. Even in the last stages of landscape formation, the Islands were in a much different circumstance than now—buried by a mile-thick sheet of ice. To begin this story we first briefly introduce the surroundings. Next we divide the rocks and surficial deposits of the San Juan Islands by ages and type, and outline the structures and processes that created the geology of today. Finally, we look at how the geology sits in the big picture of western North America.

Regional setting in the Pacific Northwest
The San Juan Islands are a relatively small speck in the overall geologic landscape of western North America. As far as immediate neighbors, north of the San Juan Islands and active during their development is the Coast Plutonic Complex (Fig. 24). This feature is a major continental arc running north 1000 miles from Washington to Alaska, that comprises many thousands of cubic miles of granitic magma intruded upward into the earth's crust from melt zones in the upper mantle. West of the San Juan Islands on Vancouver Island, is Wrangellia, a large block of rock separate in its origins from North America, also extending to Alaska, and composed mainly of 370 to 200 million year old island arc and ocean plateau rocks. It is far travelled. By mid-Jurassic time, 170 million years ago, Wrangellia was "stitched" to the continent by granites of the Coast Plutonic Complex that intruded across the junction. Wrangellia and the Coast Plutonic Complex were together, and define a northern and western boundary of the region where the bedrock geology of the San Juan Islands developed in Late Cretaceous time, ~100 - 70 Ma.

Terrane
We use the term "terrane" in referring to an assemblage of rocks the parts of which are related to one another, but the assemblage as a whole is not related to neighboring rock units. The "terrane" was broken loose from its origins and travelled to, and was faulted into place, where it is now. The older bedrock geology of the San Juan Islands is made up of several terranes of distant origins, faulted together.

Geologic units and structural assembly
Most of San Juan Islands' landscape is underlain by terranes that are large thrust sheets, termed nappes, stacked one on top of another like a deck of cards (Figs. 25-27). The nappes extend into the northwest Cascades, and including this full extent the nappe complex is referred to as the "San Juan Islands - Northwest Cascades thrust system". The nappes are enormous in breadth, up to 60 miles, but are only a mile or two thick (Fig. 25). Working out this complex geology by defining the nappes and the faults separating them has been decades in development, coming mainly from on-the-ground tracking out of rock units up and down topography through the brush and forest. University of Washington professors Peter Misch and Joe Vance made the initial large discoveries in this endeavor in the 1950s and early '60s, with their field studies in the Cascades and San Juan Islands, respectively. And the work carries on, with advances in recent years by students and faculty of Western Washington University and the University of Washington, and personnel of the US Geological Survey and the Washington Division of Geology and Earth Resources.

The lowermost nappe is made up of the Turtleback Complex, the plutonic basement of an island arc, and the East Sound Group, its volcanic cover. This arc was long lived from about 500 - 260 million years ago. Next upward in the nappe pile is a 280 -190 million year old grouping of ocean floor rocks, named the Orcas Chert and Deadman Bay Volcanics. The next higher nappe is also of ocean origin and is termed the Ocean Floor Assemblage, 170 -120 million years old. And the uppermost nappe is

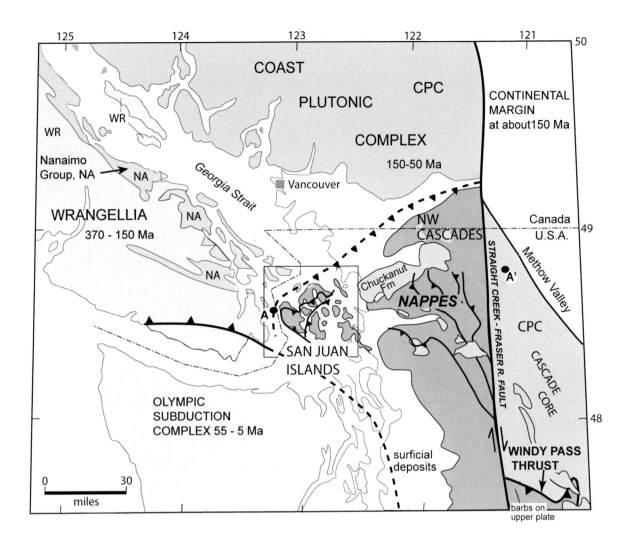

**Fig. 24 Regional maps.**
*Above:* **The major tectonic elements of the Pacific Northwest as at present. The San Juan Islands thrust sheets (nappes) are continuous into the northwest Cascades. To the east, the nappes are cut off by the Straight Creek-Fraser River strike-slip fault, active about 45 million years ago. A—A' spots mark the cross section line of Fig. 25.**

*Right:* **Regional geology prior to offset of 100 miles on the Straight Creek-Fraser River fault. The eastern and western parts of the nappe complex are congruent. The Windy Pass thrust in the central Cascades placed the nappe system over the south end of the Coast Plutonic Complex. The nappes are also inferred to be emplaced over Wrangellia which was intruded by the Coast Plutonic Complex prior to the time of thrusting, about 90 million years ago. The marine sedimentary basin of the Nanaimo Group (NA on the map) was fed by erosion of all the surrounding major elements: the San Juan Islands and northwest Cascades nappes, the Coast Plutonic Complex and Wrangellia.**

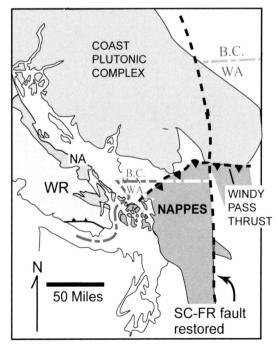

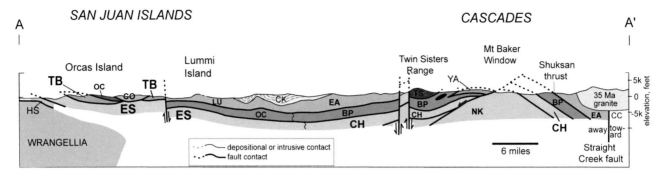

**Fig. 25** Cross section of the San Juan Islands—northwest Cascades thrust system. Section line is marked on Fig. 24. Uplift and exposure of the nappe sequence in the "Mt Baker window", mapped in the 1950s by Peter Misch of the University of Washington, defined the nappe sequence and gave us evidence of the thinness of the nappes relative to their broad extent. Rock units of the San Juan Islands are interpreted to be correlative with rocks in the Cascades that are similar and are located in the same structural position. Symbols: BP = Bell Pass Mélange, CC= crystalline core, CH = Chilliwack group, CK = Chuckanut Formation, CO = Constitution Formation, EA = Easton Metamorphic Suite, ES = East Sound Group, HS = Haro and Spieden Formations, LU = Lummi Formation, NK = Nooksack Formation, OC = Orcas Chert and Deadman Bay unit, TB = Turtleback Complex, TS = Twin Sisters dunite, YA = Yellow Aster Complex.

another island arc, 170 - 120 million years old, named the Fidalgo ophiolite. These nappes extend into the northwest Cascades (Fig. 25) and are well exposed in alpine areas, as at world-famous Mt Shuksan.

The nappes are not of North American origin; through the dynamics of plate tectonics, the nappes have come a great distance, some more than 1000 miles. How San Juan Islands nappes became positioned where they are now, considering both their long distance travel and their final emplacement, is a story of timely significance to much current geologic research. Critical questions

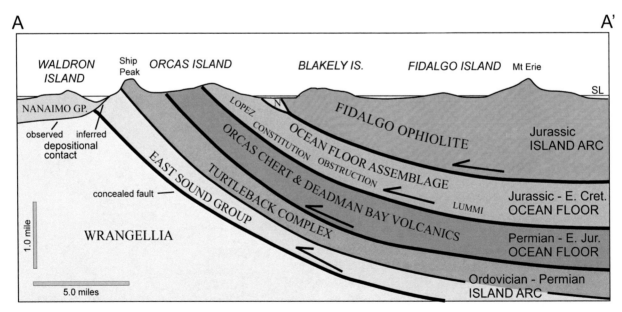

**Fig. 26** Vertical cross section along A-A' on Fig. 27, showing the San Juan Islands nappe sequence and its inferred relation to the Nanaimo Group and Wrangellia. Note the large vertical scale compared to the horizontal scale—topographic relief and thickness of geologic units are enhanced. Thrust sheets are schematic.

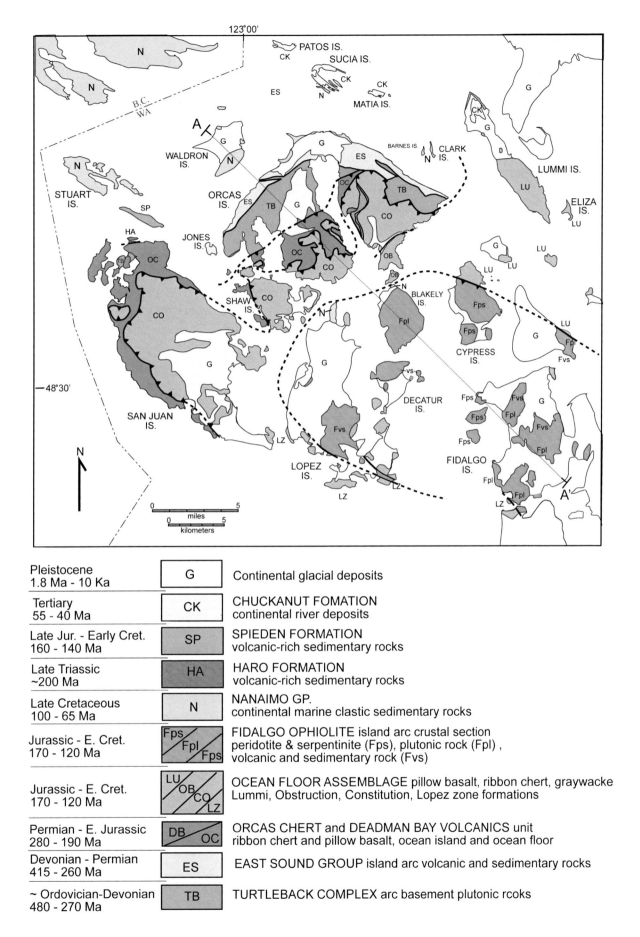

Fig. 27. Geologic map of the San Juan Islands, compiled from numerous sources. See References. A vertical section along A-A' is illustrated in Fig. 26.

are: What processes of continental growth produced these great thrust faults? What is the direction of thrusting, the homeland of the rocks, and relation of the nappes to Wrangellia, the Coast Plutonic Complex and overlying sediments of the Nanaimo Group? These questions that relate to the overall picture of continental margin tectonics in this region are addressed more fully in Chapter 12.

As a start to understanding the emplacement of the nappes, it is useful to mentally restore displacement on the Straight Creek - Fraser River fault, which slid the San Juan Islands and connected rocks to the north approximately 100 miles about 45 million years ago (Fig. 24). With this reconstruction, we see that the San Juan Islands - northwest Cascades thrust system lines up with similar rocks in the central Cascades. Near the town of Leavenworth in the Central Cascades, Bob Miller working on his University of Washington PhD thesis mapped the Windy Pass thrust (Fig. 24), a critical finding that shows that the nappe pile is faulted over the Coast Plutonic Complex. The fault is dated at 93 million years based on uranium-lead ages of cross-cutting and faulted plutons in the Windy Pass area. We don't see it but assume there is a similar thrust fault under water somewhere between the San Juan Islands and Wrangellia. Thus, we envisage a broad stack of thin nappes extending from the central Cascades to the San Juan Islands thrust over the south end of a landmass of co-joined Wrangellia and the Coast Plutonic Complex. From uranium-lead and fossil dating, thrusting began perhaps 100 million years ago and occurred episodically emplacing different nappes until at least 75 million years ago.

North of the thrust system, and in part occurring within it, are marine sedimentary rocks named the Nanaimo Group. Fossils of the Nanaimo Group and radiometric ages indicate deposition from about 95 - 65 million years ago. The Nanaimo Group contains sediment from the surrounding geology: Wrangellia, the San Juan Islands nappes, and the Coast Plutonic Complex. Thus it accumulated in a basin surrounded by these geologic features, which were in place approximately as we see them today. The Nanaimo sedimentary strata consist of lithified (compacted, cemented) mud, sand and gravel. The Nanaimo Group came into existence during thrusting in the San Juan Islands, and bearing sediments distinctive of the individual nappes helps us understand the timing of nappe emplacement.

The youngest rock unit comprising the San Juan Islands architecture is the Chuckanut Formation. This rock, like the Nanaimo Group, is derived from mud, sand, and gravel. But the Chuckanut sediments yield abundant evidence of a continental origin in the form of preserved animal tracks, terrestrial plant fossils, and sedimentary features that indicate deposition by streams and rivers. As the sediments accumulated over a period some 55 - 45 million years ago, the apparently fault-bounded basin floor subsided hugely, making room for a formation up to 5 miles thick over parts of the San Juan Islands and Cascades.

Surficial geology of the Islands, developed atop and after the bedrock geology, is largely represented by sedimentary deposits and rock sculpture caused by a lobe of the Cordilleran continental glacier—a one mile thick sheet of ice crossing the Islands during an interval 18 - 16 thousand years ago.

<u>The North American Cordillera</u>
Terranes of the San Juan Islands nappes have relatives elsewhere in North America, occurring in the broad expanse of materials added to the western margin of the continent since late Precambrian time (Fig. 28).

This region is known as the "North American Cordillera". The beginning of the story goes back about a billion years when virtually all of earth's continents were glommed together in a supercontinent we call "Rodinia". Ancient North America, named "Laurentia", was the core of this assemblage of continents. On the eastern margin of Laurentia was Europe. Attached on Laurentia's western margin were Australia, Antarctica, possibly Siberia, and parts of Asia. As breakup occurred about 800—600 million years ago, the ancestral Atlantic ocean opened on the east coast of Laurentia, and along the

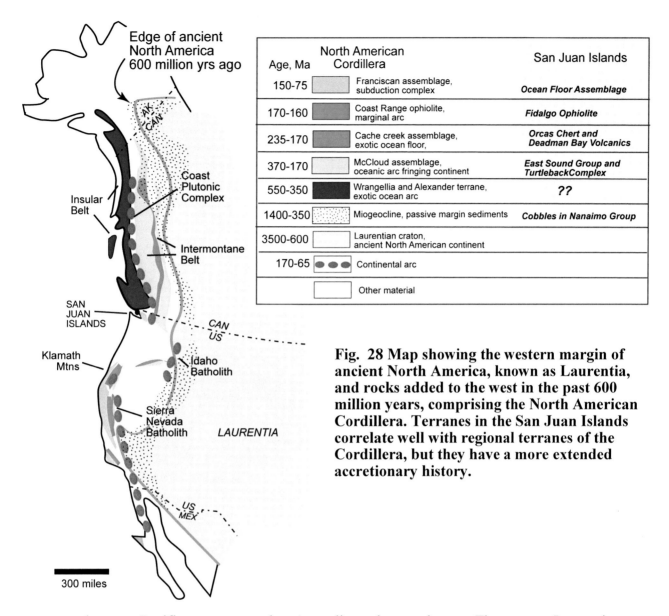

Fig. 28 Map showing the western margin of ancient North America, known as Laurentia, and rocks added to the west in the past 600 million years, comprising the North American Cordillera. Terranes in the San Juan Islands correlate well with regional terranes of the Cordillera, but they have a more extended accretionary history.

west coast the proto Pacific ocean opened as Australia et al. moved away. The western Laurentian margin was what we call "passive", no plate convergence and no arcs. Local sediments filled extensional basins, some with zircon age signatures of the now long-gone terranes. Over a broad scale along western Laurentia, in the region of the present day Rocky Mountains, sediments of very mature types including quartz sandstone, were laid down by slow moving rivers and in shallow seas on low-relief coastal areas—the "miogeocline" of Fig. 28.

This peaceful tectonic environment lasted from about 600 –350 million years ago. After that time intermittent episodes of contraction are recorded; the margin became "active", and we begin to see evidence of island arcs fringing the continent. Beginning at about 200 million years ago, arcs and ocean crust were accreted on a broad scale and thick prisms of sediment were laid down, all adding great tracts of land to the continent. In sum, the North American Cordillera consists of late Precambrian and early Paleozoic passive margin sedimentary rocks to the east in the Rocky Mountains, and belts of accreted ocean crust, island arcs, and thick accumulations of greywacke in the west. Continental arcs and local sedimentary basins overprint the accreted terranes. These belts of rock and their linkage to the San Juan Islands are illustrated in Fig. 28, and are further considered in subsequent chapters.

# CHAPTER 5
# *TURTLEBACK COMPLEX*

The oldest rock on the San Juan Islands, yielding uranium-lead ages of zircons in the range of 350-500 Ma, occurs as mixtures of granite and gabbro that take us into the realm of deep crustal magma chambers. The rock is named the Turtleback Complex after its broad extent underlying Turtleback Mountain on Orcas Island (Fig. 29). Good exposures occur in coastal areas of Orcas Island, and in the public lands of the San Juan County Trust on Turtleback Mountain, and in Moran State Park (Figs. 31-33).

**Fig. 29 View of the Turtleback Mountain from the south, looking up West Sound.**

The Turtleback Complex is considered from study of rock types and chemical composition to represent the roots of an arc system (Figs. 4, 30). In this setting, magma formed by melting in the earth's mantle. The magma intruded upward, then paused near the base of the earth's crust forming a magma chamber (Fig. 9); there it fractionated by crystal settling to produce many plutonic rock types marked by various proportions of dark and light crystals (Fig. 31). These derivative magma types were

**Fig. 30 Plate tectonic model for the origin of the Turtleback Complex (TB) and East Sound Group (ES) as an island arc. The arc is fed by magma generated by melting in the upper mantle over a subduction zone. In rising, some magma stalls and crystallizes at depth forming the coarse-grained plutonic rock of the Turtleback Complex. Other magma is intruded near the surface, or erupts, and is volcanic rock. The island arc moves; it is drawn toward its magma source over the subduction zone, which steps back. The converging and subducting ocean plate is constantly sinking down and pulling back while it is diving into the subduction zone. New ocean crust is formed in a spreading zone behind the arc. By this mechanism, arcs moving a few inches a year can travel great distances in the oceanic realm.**

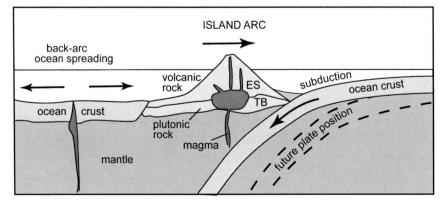

Fig. 31 Layered cumulate gabbro. The layers formed by accumulation of crystals along the flanks and bottom of a magma chamber. Compositional variations in the magma chamber produced different crops of crystals; dark layers are richer pyroxene, light layers are rich in feldspar. The folds indicate slumping of crystal mush in a magma chamber. This is a coastal exposure along the east shore of East Sound, near the head of the Sound.

Fig. 32 Igneous injection complex; dikes of varying composition are intruded into one another. Here, light-colored granite is intruded by basalt. Coastal exposure at West Sound.

Fig. 33 Lighter colored diorite is intrusive into older gabbro. The locality is at Deer Harbor, along the a coast few 100 feet south of the marina.

then intruded up into the crust, one against another, forming an injection zone of multiple separate light and dark colored dikes (Figs. 32-33).

The Turtleback Complex is a relative of the East Sound Group, which is mainly volcanic rock, described in the next chapter. The two rock units are closely associated, overlap in age, and are of arc origin. We find fine-grained dike rocks in the Turtleback Complex that are similar to volcanic rocks of the East Sound Group and could be feeders to volcanoes. So, we can think of the Turtleback complex as the exposed basement, the magma works, and the East Sound as overlying volcanoes. This scenario is complicated, however, by the position on Orcas Island of the Turtleback Complex *on top* of the East Sound Group (Fig. 26). We must conclude either that the original section was flopped upside down, or much more likely, that a fault has pushed the Turtleback up over the East Sound Group.

Where did this ancient arc originate? The younger ages of the Turtleback Complex, 350-400 Ma, and those of the East Sound Group match well with ages of other accreted parts of western North America (Fig. 28). As a group, these old arc rocks are thought, based mainly on fossil distributions, to have formed in an ocean setting marginal to western North America, ranging from the northern Sierras in California to the central part of British Columbia. But, the older ages of the Turtleback Complex, 400-500 Ma (Figs. 34, 35), do not find western North American counterparts. These rogue ages match with the Appalachians in eastern North America, and even farther afield to northwestern Europe! These findings, if they last the test of further study, require plate tectonic transport for the older parts of the Turtleback Complex of some 3000 miles around the northern or southern flanks of Laurentia.

**Fig. 34 Michael Yeaman collecting a sample of Turtleback Complex for zircon age-analysis. This sample comes from a patch of bare rock exposed under an uprooted tree along the path near the summit of Ship Peak. About 20 lbs. were obtained and subsequently crushed in the lab. The crushed material was placer-mined on a water table to concentrate the heavy zircon. Further processing of the concentrate included magnetic and heavy liquid separations. From the original rock, the final zircon separate was less than a pencil eraser in volume.**

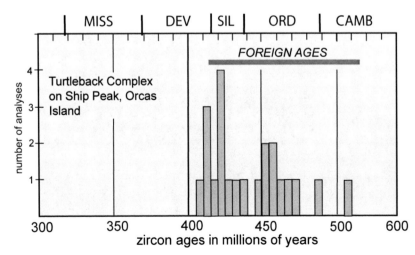

**Fig. 35 Spot ages on zircon crystals from the sample of Turtleback Complex collected by Michael Yeaman, above. Most ages fall outside the known limits of igneous rocks in Western North America, but do match ages in the Appalachians and Scandinavian Calidonide Mountains.**

# CHAPTER 6
# *EAST SOUND GROUP*

From an aesthetic standpoint, the East Sound Group is mostly ugly. At the type locality in the Waterfront Park at the head of East Sound, rocks are dark, slightly greenish, and generally massive; in other words, inscrutable. For better exposures, take a cruise around the Jones Island State Park where wave-washed coastal outcrops more clearly tell a story of rock history.

Most of the East Sound rocks are volcanic (Fig. 36). Some occur as lava flows, but generally they are explosive deposits, termed pyroclastic, ranging from coarse breccia to tuffaceous sands. Marine limestones, formed as reefs, are locally intruded by volcanic rocks (Fig. 37), and importantly provide datable fossils. In the past, limestone in the East Sound Group had value as the raw material for cement, and the old quarries can still be found. Sequences of interbedded greywacke sandstone and siltstone also occur in the East Sound Group (Figs. 38, 40). These sediments apparently represent turbidity deposits shed off a continental margin, perhaps triggered by earthquakes (Fig. 22).

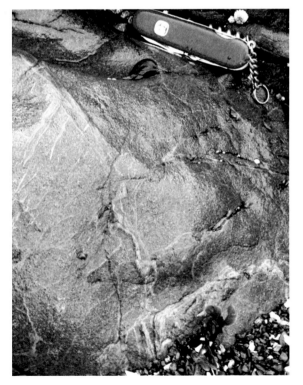
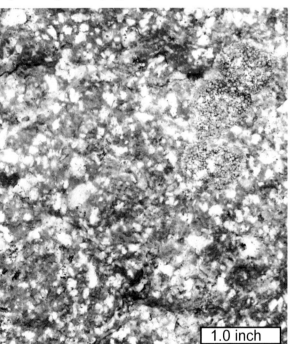

**Fig. 36 Volcanic rock types of the East Sound Group, all in coastal outcrops around Jones Island. The *lower left* picture shows greenstone. This rock is a basalt or andesite that has been slightly recrystallized by metamorphism to produce chlorite and other low-grade green minerals from original pyroxene and amphibole in the igneous rock. The *upper left* photo is of volcanic breccia formed by an igneous explosion in which lava fragments and hot gases were blasted out during volcanic eruption. The *lower right* picture is andesite with a fine-grained groundmass and large white feldspar crystals.**

The East Sound Group represents an ancient island arc. As described in the previous chapter, plutonic rocks of the Turtleback Complex are interpreted as the basement to the East Sound volcanoes. The age span of East Sound rocks is about 415 - 260 Ma, whereas the Turtleback plutonic rocks range from ~500 to 350 Ma.

So, here we have island arc volcanic rocks with the basement magma chambers and feeding dikes exposed, roaming at large in the ocean for on the order of 150 million years. Where was this arc? Fossils in the limestone, especially those in the age range of 250 - 300 million years old, suggest a co-mingling of oceanic and western North American continental organisms, indicating proximity of these oceanic arc rocks to the continent. Also supporting this concept, the turbidity deposits in the East Sound Group contain abundant zircon crystals that match ages in the ancient North American continent (Fig. 39). The East Sound arc can be considered to be "fringing" the western North American continent, in a way similar to fringing arcs in the western Pacific today (Fig. 5).

How does the East Sound Group relate to other rocks of the region of similar age? Early studies in the San Juan Islands grouped the rocks we now call East Sound Group together with the Orcas Chert and Deadman Bay Volcanics, but as Mark Brandon and others of the University of Washington noted, the fossils indicate very different origins of these two rock units; they are not related. Another thought is

**Fig. 37 Limestone is interbedded with sandstone in the East Sound Group on the east side of Jones Island. Above Liz Schermer (taking notes), is a dark mass of igneous rock intrusive into the middle of the limestone, clearly indicating the origin of the limestone in a volcanic setting. This limestone is the oldest known sediment in the East Sound Group, dated by Conodont fossils as earliest Devonian, about 415 million years old. Conodonts are tiny teeth or jaw bones of eel-like animals, the soft parts of which are long gone.**

Fig. 38 Broken and bent bedded greywacke sandstone and siltstone of the East Sound Group on the south end of O'Neal Island. Sedimentary features indicate deposition by turbid flows across the sea floor (Fig. 22); thus, the rocks are termed turbidites. These sediments are associated with volcanic arc rocks typical of the East Sound Group, and some of the zircons in these beds give 300-400 Ma ages (Fig. 39) consistent with fossil ages of limestones interbedded within the East Sound Group arc. But most zircon sand grains in these greywacke beds are much older than the associated volcanic arc rocks and match ages in the ancestral western North American continent (Laurentia, Fig. 28). The old zircon ages indicate an origin of the sediment by erosion of the North American continent, suggesting that the East Sound Group arc formed not so far away.

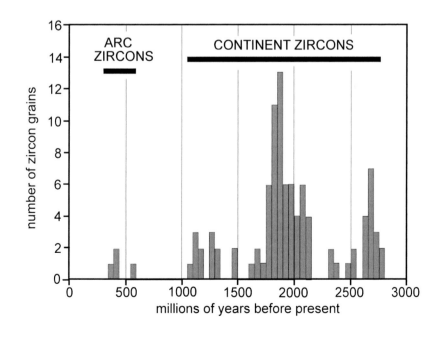

Fig. 39 Histogram showing ages of zircon sand grains in greywacke of O'Neal Island. The abundance of continent-derived grains indicates origin of the East Sound arc not far from western North America.

that the older bedrock of Vancouver Island may correlate with the East Sound Group. If this were true, then the San Juan Islands would be considered part of the exotic terrane Wrangellia. However, again, fossils distinguish East Sound rocks from Wrangellia. Instead, the paleontology points to correlation of the East Sound Group with island arc rocks in the North Cascades, there termed the Chilliwack Group (Fig. 25).

More broadly, the East Sound Group is part of a great extent of similar accreted arc rocks in the Western Cordillera generally referred to as the McCloud assemblage, in reference to distinct fossils occurring the northern Sierras, the eastern Klamath Mountains of California and Oregon, the Blue Mountains of eastern Oregon, and central British Columbia (Fig. 28). The fossils that link the separate occurrences, also loosely tie the rocks to western North America which has some related species. A strong population of zircons older than 1000 Ma also ties the East Sound Group to the ancestral western North American continent (Fig. 39). We imagine this expanse of rocks to have lain as an arc system some 1000s of miles long, outboard of the continent, and lasting for at least a couple of hundred million years. Eventually the arc was added to the continent, accreted, at about 170 million years ago.

**Fig. 40 More turbidites (compare with Fig. 38); these occur along the northwest shore of Orcas Island. The light colored beds are sandstone and the darker ones siltstone. The flow of turbid sediment across the sea floor, which has settled out to make these light and dark beds, was likely triggered by earthquakes shaking sediment loose from the coastal margin (Chapter 3, Fig. 22). To the extent that this explanation holds true, each couplet of light and dark sediment represents an earthquake. Much seismic activity is represented here.**

# CHAPTER 7
# *ORCAS CHERT AND DEADMAN BAY VOLCANICS*

A truly oceanic assemblage of rocks is presented by the Orcas Chert and the related Deadman Bay Volcanics. Chert, reef-formed limestone, and ocean floor pillow basalt make up the unit. These rocks record a long-lived and broad segment of ancient ocean plate, with origins in the ancestral western Pacific, and now disappeared under the western flank of North America except for accreted scraps we see in the San Juan Islands and locally elsewhere in the Cordillera.

The Orcas Chert unit is mostly composed of a rock type known as "ribbon chert" because of its distinctive layering (Fig. 41). Light-colored, hard and weather-resistant quartz layers 1-2 inches thick alternate with somewhat thinner, recessive clay-rich beds—giving the "ribbon" appearance (Chapter 3). In microscope view, the quartzose layers can be seen to have some tiny (0.1 - 0.5 mm) relict fossils identified as radiolaria, silica skeletons of single cell plankton. The chert is hardened plankton ooze that accumulated on the deep ocean floor. The clay-rich interbeds represent a component of airborne dust and silt wafted over the ocean. How the layering developed is not easily understood. Instead of the layers forming by pulses of different sediment, as in bedded shales and sandstones, the radiolaria vs. silt layers more likely formed during reconstitution and compaction which caused un-mixing of the silica-rich and clay-rich components of original soft sediment ooze.

**Fig. 41 Ribbon chert of the Orcas Chert unit. Good outcrops of this rock are seen along the east shore of Cascade Bay near the Rosario Resort (Chapter 14). The chert layers are bent and broken, testimony to travels of this rock as ocean crust that collided with, and was thrust far under, the continental margin.**

**Fig. 42 White limestone beds interleaved with sheared chert and volcanic rock of the Orcas Chert formation in the Dolphin Bay Quarry on Orcas Island. This rock is actively mined by Island Excavating Inc. for landscaping and road materials.**

Radiolarian fossils evolved with time and provide excellent age markers for these cherts. Even within a single mapped unit of ribbon chert, age differences can be determined; typically the rate of formation discovered for these deposits is 3-5 feet per million years (Chapter 3). Applying this age measure to the Orcas Chert, which is up to 1500 feet thick, has problems in that the measured thickness results not only from the original chert deposit but also is a result of folding and faulting which bunch-up the section. The age range of the fossil radiolaria in the Orcas Chert is about 240 - 180 Ma (Triassic to Early Jurassic).

Interlayered within the ribbon cherts are subordinate basalts and limestone beds. Over decades these limestone beds, which are up to 100 feet or more thick, have been quarried. In the early days the limestone was fed into kilns and reduced to lime. More recently, as in the active Dolphin Bay quarry on Orcas Island (Fig. 42), limestone is mined for construction and landscaping material. The limestone is more accurately termed marble as it is a metamorphic rock made up of the high pressure form of $CaCO_3$, aragonite, crystallized in a subduction zone (Figs. 15, 16).

The Deadman Bay Volcanics (Figs. 43, 44) are spatially associated with and are somewhat older than the Orcas Chert, and are probably part of the same ocean crustal section. But faults separate the units, leaving some uncertainty about the connection. The Deadman Bay Volcanics consist mostly of pillow basalt. The pillows developed during submarine eruption and flow of basalt magma (a process spectacularly filmed in Hawaii where Kilauea volcano erupts into the ocean). An ocean island origin of the basalts, similar to the Hawaiian Islands, is indicated by their chemical composition (Fig. 14).

Interleaved with the basalt in the Deadman Bay Volcanics, and in interstices between pillows, are shallow-water limestones containing a very interesting fossil marker—fusulinids of exotic origin (Fig. 45). These animals are single-celled, like the radiolaria, but differ by being calcareous not siliceous, and in living on the sea floor as opposed to the planktonic (floating, swimming) radiolaria which could range widely in ancestral oceans. The fusulinids of the Deadman Bay Formation are of special interest to tectonics because they are insular in range (they did not crawl across ocean basins); they are unlike fusulinids of similar age in ancient North America, and are also unlike the comparably aged fusulinids of the East Sound Group. As recognized in the early 1970s by Ted Danner of the University of British Columbia, the Deadman Bay fusulinids match with rocks formed in the ancient Tethys Sea in the far western proto-Pacific, thousands of miles from ancient North America (Chapter 12). The age of these exotic fusulinids is about 280-260 Ma (Permian). Plate tectonics brought these fossils to our shores.

The Deadman Bay Volcanics are somewhat older than the Orcas Chert and likely formed the basement to the chert. Applying the details of formation that we learn from the rock types and fossils, these rocks define an exotic oceanic terrane that was born on the western side of the ancient Pacific Ocean. It began as ocean-island volcanoes over a mantle hotspot (Chapter 1). It subsequently moved off the hotspot and evolved into a long-lived ocean plate that travelled some thousands of miles eastward to be finally accreted to North America. The combined time span of these units ranges from about 280 to 180 Ma, meaning about 100 million years of ocean residence. At a typical plate motion rate of 4 inches/year, the Lime Kiln Point rocks could have travelled 6000 miles, or possibly twice that far at a faster but feasible speed. The distance from the Tethys realm to western North America across the ancient Pacific Ocean was perhaps as much as 10,000 miles considering that the Atlantic Ocean at this time was closed (Chapter 12).

**Fig. 43 Lime Kiln Point lighthouse and rock mixture of ocean island basalt (brown) and ingrown reef limestone (grey) all in the Deadman Bay Volcanics rock unit. This limestone is host to exotic fusulinid fossils (Fig. 45).**

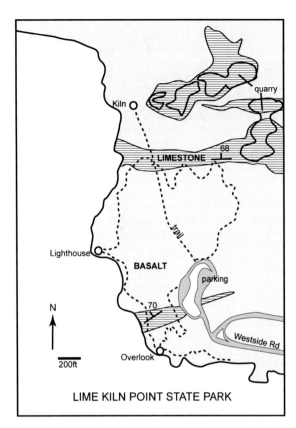

Fig. 44 Map of Lime Kiln Point State Park. Bedrock here is Deadman Bay Volcanics.

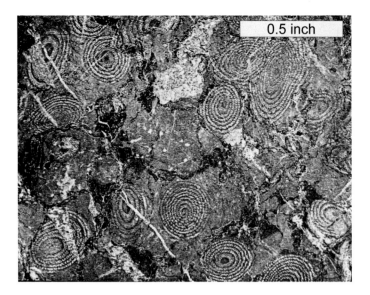

Fig. 45 Microscope photo of Permian fusulinids, identified as a species of Yabeina, occurring in limestone at Lime Kiln Point. The fossils are exposed on a cut slab, and show in cross-section a concentric shell structure. These animals are a type of foraminifera. The known range of this species in Permian rocks is in the Tethys Sea, near the coast of Asia on the western side of the ancestral Pacific Ocean (Chapter 12). The fusulinids lived on the sea floor (benthic), and were not widely dispersed as are floating (pelagic) plankton. Therefore their existence here is evidence that the limestone is very far travelled. Photo from Danner (1966).

Fig. 46 Kiln at Lime Kiln Point. For 90 years beginning in 1860, limestone was quarried on the hillside above, largely by pick and shovel. The limestone, $CaCO_3$, was cooked in this furnace over a wood fire to produce lime, $CaO$. Ships docking in the straits below the kiln, and alternatively at the more protected Roche Harbor, carried the lime to many west coast ports.

An exotic rock component of the Orcas Chert, adding a dimension of mystery to the geologic history, is the Garrison Schist. This rock is of oceanic origin, consisting mainly of dark green metamorphosed basalt and well recrystallized ribbon chert. Metamorphic amphibole dated by K-Ar gives ages in the range of 240-280 Ma, somewhat older but overlapping the sedimentary age of the Orcas Chert. The rock occurs as lenses within Orcas Chert, either as fault slices or possibly as large slabs that have slid into the Orcas Chert sedimentary basin. The rock is not easily distinguished from the Orcas chert in outcrop, but can be found on San Juan Island along the coast in American Camp National Historic Park. Outcrop locations are given in a paper by Brandon and others (1988).

# CHAPTER 8
# OCEAN FLOOR ASSEMBLAGE

The package of rocks that we call the "Ocean Floor Assemblage" tells a plate tectonic story of formation of ocean crust and its transport to a landmass (Chapter 1, Fig. 47). Basalt magma erupted at a ridge or hot spot; the magma was quenched to form basalt rock, and thereby became a newly formed part of an oceanic plate. This plate moved away from the magma vents and across an ocean basin where with time and distance it became covered by cherts, formed by accumulation of radiolaria settling down from the ocean above. Eventually the plate, now basalt covered by chert, moved into the vicinity of a highland shedding off great amounts of sand and silt sediment. The telltale assemblage is pillow basalt, overlain by ribbon chert, overlain by greywacke sand- and siltstone. The pillow basalts have a distinctive chemical composition separating them from island arc rocks (Fig.14).

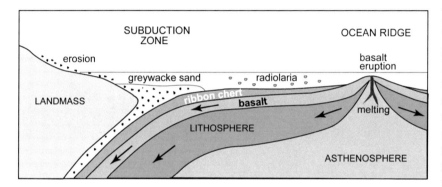

**Fig. 47 Schematic drawing of the tectonic evolution of ocean floor rocks of the San Juan Islands. The process starts at the ocean ridge, where basalt erupts from a melt source in the mantle below. The basalt cools and crystallizes, and then the basalt and underlying mantle slide away from the ridge on the slippery asthenosphere (Fig. 4). As the basalt travels away from the ridge, radiolarian plankton slowly accumulate on the sea floor forming an ooze, that is eventually compacted into chert—the longer the distance travelled, the thicker the chert. As the ocean crust nears land, eroded sediment is deposited on top, as greywacke sand and silt. Thus, in outcrops in the San Juan Islands, we see a succession of layers from the bottom up of: basalt - chert - greywacke. Complexity in the history is added by evidence in the San Juan Islands of the oceanic rocks going down the subduction zone (presence of aragonite, Chapter 3) and then coming back up.**

A vast amount of the rock exposed in the San Juan Islands is greywacke, a dark sandstone not easily analyzed without a microscope. Some of this rock belongs to the Paleozoic East Sound Group, as described earlier (Chapter 6). But most of the greywacke is a younger formation, dated as Late Jurassic to Cretaceous, 150 - 120 Ma. There is not much to go on for geologists trying to assign these greywacke outcrops to one major rock unit or another, a basic step to understanding the structural makeup of the Islands. The current best idea is to distinguish the greywackes associated with island arc volcanic rocks from those connected to ocean floor volcanic rocks, an approach conceived by USGS geologist Clark Blake and others. Thus, we distinguish the greywackes that are associated with igneous rocks of arc origin in the Fidalgo Ophiolite (next chapter) from the remaining greywackes that overlie ocean ridge and ocean island basalts and ribbon cherts.

A challenge remains for understanding the mutual relations of the Ocean Floor Assemblage in different areas of the San Juan Islands. Different assemblages of the basalt-chert-greywacke sequence are distinguished on the map and cross sections (Figs. 25-27), but they may or may not represent contiguous regions of ocean floor. Some assemblages are perhaps of different age or are much farther travelled than others or have a closer relationship to arc rocks. In the structural model shown in Fig. 26, all components of the ocean floor assemblage are loosely grouped into one nappe, but here also we have uncertainty and are not sure whether these parts were emplaced in the San Juan Islands as a single nappe or arrived separately.

Lopez Structural Complex

Ocean floor rocks are beautifully exposed as coherent chunks in an otherwise structurally mixed zone we call the Lopez Structural Complex, lying along the southern edge of Lopez Island and the south end of Fidalgo Island (Fig. 27). A great place to see the layered sequence of ocean floor rocks is at Rosario Head in Deception Pass State Park (Figs. 48-50). Here, the geology tells a fairly complete plate tectonic story of birth and accretion of ocean crust. Pillow basalts are well displayed on the cliff face of the Head. Overlying the pillows is ribbon chert exposed over the top and down the sides of the Head. On the flanks of the Head dark siltstone (argillite) overlies the chert.

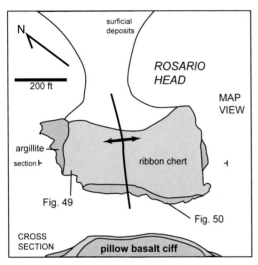

**Fig. 48 Geologic relations at Rosario Head in Deception Pass State Park.** An oceanic stratigraphic succession is displayed from the bottom up of pillow basalt, ribbon chert and argillite (compacted silt and mud). This sequence represents evolution of the crust from its initial formation by basalt magmatism at an oceanic spreading center or hot spot; to residence and accumulation of radiolarian ooze during plate transit across quiet waters of the deep ocean; and finally to arrival of the plate into the vicinity of a highland with a source of silt and mud—possibly an island arc or the continental margin (Fig. 47).

**Fig. 49 Ribbon chert.** Silica-rich white layers stand out in relief against thinner, recessive, clay-rich dark layers. This rock forms from deep ocean sediment rich in siliceous plankton—radiolarian ooze.

**Fig. 50 Pillow basalt on the Rosario Head cliff face.** Each pillow represents a blob of basalt magma that was extruded out over previously formed pillows, producing a distinctive shape with a keel structure at the pillow base where the magma came to rest over the older pillows. This outcrop can be viewed from the water, or by very careful traverse near the waterline from the south corner of the Head.

As discussed earlier (Chapter 3), ribbon chert forms by slow accumulation and compaction of radiolarian ooze on the ocean floor at a rate of about 3 - 5 feet/million years. For this deposit of about 30 feet of chert we can estimate 6-10 million years of residence on the sea floor after the plate moved off the magma source and before it moved into an area of heavy silt sedimentation, such as a trench near the coast of eroding land (Figs. 4, 22, 47). Also important is the presence of aragonite, a sign of subduction. Using these facts, we interpret that the geology first formed as ocean crust over a hot spot or ocean ridge vent system. The crustal plate was moved off the magma source by sea-floor spreading and, while accumulating chert, travelled toward a land mass. During this 6-10 million years of chert formation, the plate could have moved perhaps 25 - 40 miles at a typical velocity (4 inches/ year)

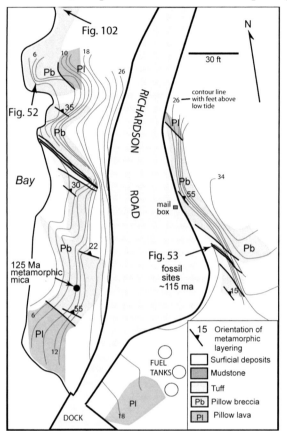

before being loaded up with silt sediment next to a land mass. Probably not long after this point, the plate was subducted, passing a depth of 12 miles down, where calcite converted to aragonite (Fig. 16). Finally the plate was uplifted, and by a great amount of erosion came to the earth's surface.

**Fig. 51 Geologic locality at the end of Richardson Road on Lopez Island. A complex history involving large vertical crustal displacements is indicated. From this locality a metamorphic age of 125 Ma was measured on mica in deformed and recrystallized basalt pillow breccia. The breccia contains the high pressure mineral aragonite, indicating burial of the rock more than 12 miles deep at that time. Also occurring at this locality is mudstone with foram fossils of about 115 Ma; these rocks are crossed by aragonite veins. A complex history emerges. Sea floor basalts were subducted and metamorphosed at 125 Ma. Elsewhere, later, mud accumulated on the sea floor together with ~115 Ma forams, and this rock was also subducted. Still later, the older volcanic rock and younger mudstone were mixed together by faulting—possibly during some part of the process of uplift out of the subduction zone.**

Fig. 52 Exposure of basalt pillows as 3-D forms. Red Swiss army knife for scale. Location on the map above.

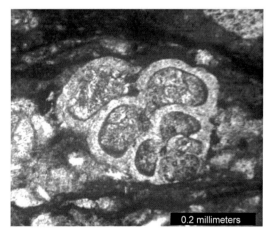

Fig. 53 Globigerina foram in mudstone, viewed through the microscope. Fossil site on the map.

Another well exposed and accessible place to see ocean floor geology in the Lopez Complex is at Richardson on Lopez Island (Fig. 51). Here, the geologic facts paint a picture of vigorous rock activity. Ocean basalt (Fig. 52) erupted on the sea floor, then went down the subduction zone to be overprinted by metamorphic aragonite and mica at 125 Ma (the mica dated by Ar geochronology). This rock was faulted at depth against mudstone with foram fossils (Fig. 53) as young as about 115 Ma. The mudstone was overprinted by veins of aragonite, indicating continued subduction zone metamorphism. Finally, uplift and erosion occurred. (Whew!) Rock exposures here show some of the best basalt pillow structures and volcanic explosion breccias on the Islands.

Constitution Formation

The Constitution Formation underlies large parts of San Juan and Orcas islands. This rock unit, as important as it is in San Juan Islands geology, is not easily analyzed because components are displaced and mixed by fault slicing, and because the massive nature of sedimentary outcrops typically does not show bedding which would allow structural insights. But, we do have intriguing clues as to its origin. The main component of the unit is greywacke sandstone and siltstone rich in andesitic volcanic grains (Fig. 54), indicating erosion from an arc. Ocean floor basalts and ribbon cherts are interlayered with the sandstone. Sand grains and cobbles include, besides arc detritus, metamorphic rocks and minerals that have been uplifted and eroded from a subduction zone. Cobbles are rounded, indicating shaping by streams or on a beach above sea level.

**Fig. 54 Outcrops of the Constitution Formation can be seen in a road-cut on Orcas Island at the waiting area for the ferry. Angular green andesite volcanic chunks are set in a dark groundmass of fragmental volcanic material. Veins are quartz and aragonite, the latter indicating high pressure metamorphism.**

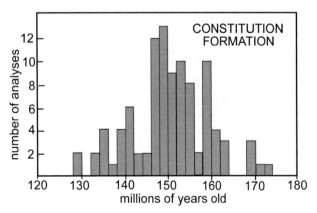

**Fig. 55 Ages of zircon sand grains in the Constitution Formation (sample site in Chapter 14). The ages show that the Formation is younger than 130-135 million years and that most of the sand, which is of arc origin, came from volcanoes 147-160 million years old.**

A reasonable thought is that the source of the arc and metamorphic clastic material could be the underlying nappes of Turtleback, East Sound, and Orcas-Deadman Bay units, which contain all these materials. Fortunately we have data to test this hypothesis from U-Pb ages of zircon sand grains (Fig. 55). No ages older than 180 million years are found, ruling out these older formations as the source. Most of the sand is about 150 +/- 4 million years old; this is the age of the prominent arc source. A maximum age of the formation is established by the youngest sand grains: two at 129 Ma, and six at 135 +/-1 Ma. Conservatively, we can say the rocks are younger than 135 Ma, probably even younger than 129 Ma, but we like to have more than two analyses to be sure it is not an aberrant number.

A further feature of Constitution Formation, as with all other units in the Ocean Floor Assemblage, is that subduction zone metamorphic minerals overprint all parts of the unit. Putting these facts together, the Constitution Formation appears to have formed by accumulation of sediment eroded from a nearby landmass made up of rocks from: 1) an uplifted subduction zone, and 2) a 150 Ma arc. The sediments were swept onto the deep ocean floor where there were intervals of basalt eruption and accumulation of siliceous ooze to form ribbon cherts. Deposition of sediments occurred until at least 135 Ma; after that the whole unit went down a subduction zone and came back up. This is not a totally satisfying conclusion in that it requires a sea-floor spreading center adjacent to an arc, and gaps of sand and silt sedimentation on the sea floor long enough to allow chert formation. We need more work here!

Obstruction Formation

The Obstruction Formation is well displayed at the Obstruction Pass State Park (Fig. 56) and along the shores of Obstruction Island. The rock consists of well bedded gray turbidites rich in chert fragments. It is locally coarse enough to be called conglomerate. The Obstruction Formation is similar to other parts of the Ocean Floor Assemblage in the occurrence of bedded greywackes and aragonite metamorphism, but differs in that basalts and cherts have not been found in this unit. Dated zircon sand grains indicate a depositional age younger than 120 Ma (not yet published data), apparently placing the Obstruction rocks in a younger plate tectonic setting than the other ocean floor rocks.

**Fig. 56 The Obstruction Formation at Obstruction Pass State Park, Orcas Island. Well bedded greywacke turbidite sandstone is rich in chert and volcanic fragments.**

Lummi Formation

The Lummi Formation is a part of the ocean floor assemblage that underlies much of the eastern San Juan Islands. This rock displays the same layered sequence of basalt - chert - silt/greywacke sandstone as at Rosario Head and is also affected by high pressure aragonite metamorphism. Here the greywacke sedimentary section greatly dominates. A thickness of more than 2500 feet is exposed on the western side of Lummi Island (Fig. 57). These sediments, ranging from gray sandstone to siltstone to dark shale, are interbedded on a scale of inches. They are commonly graded by grain size in a single bed from light to dark - base to top, in a fining upward sequence (Fig. 58). Deposition is understood to occur by sediment-laden turbidity currents that flowed down the flanks and across a marine basin, slowing and dropping coarser, then finer materials (Fig. 22). Sand grains are mostly volcanic rock and chert (Fig. 59).

**Fig. 57 Lummi Island displays a classic ocean floor assemblage. On the west side shown here is a very thick section (> 2500 feet) of volcanic-rich greywacke sandstone. Associated with the sandstone is ocean floor chert, accessed by boat along the shore near the right of the photo. Not exposed in this view but on the other side of the island, is ocean floor pillow basalt. Sandstone beds (ss) dip (are tilted) to the right in this view, thus the thickness of layers is measured normal to the dip.**

**Fig. 58 Bedded greywacke turbidites near the base of Lummi Mountain on the left horizon in the photo above. Quartz veins crossing the bedding crystalized from hot water (T>400F) penetrating along cracks during metamorphism. Quarter for scale.**

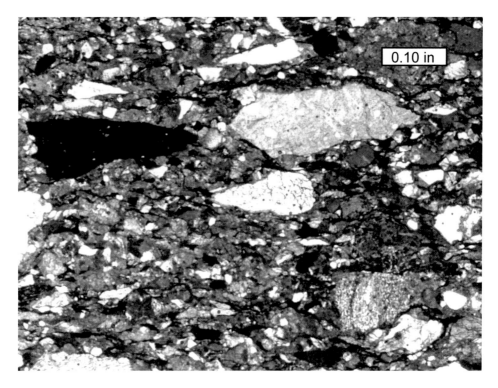

**Fig 59 Microscope view of Lummi greywacke. Most of the sand grains are green metamorphosed volcanic fragments. Large grains are light colored chert and black argillite. A slight flattening fabric has formed by metamorphic compaction and grain recrystallization.**

The Lummi Formation bears many similarities to, and seems to grade into, the Easton Suite of the Cascades (Fig. 25), which is approximately the same age and also consists of ocean floor rocks. The Easton Suite is generally more metamorphosed than San Juan Island rocks. But on North Cone and Eliza islands (Fig. 60), we see metamorphic folds and fabric (foliation), that resembles structure in Cascades rocks.

Mystery Source
Where did this great volume of greywacke sediment in the Ocean Floor Assemblage come from? Or put another way, what landmass did the ocean floor plate run into?

Microscope study shows an abundance of volcanic sand grains. And, analysis of the ages of this sand detritus by U-Pb zircon dating shows a near absence of ages older than about 170 Ma, thus there was no input from older rocks in the San Juan Islands, or from the old North American continent of that time.

A great concentration of ages is in the range of 145-150 Ma. The landmass was volcanic and apparently not part of ancient North America. A good bet is that it was an island arc somewhere out in the ocean, and not connected to the continent. Modern analogues lie in the present day western Pacific, discussed earlier (Fig. 5).

Additional useful information comes from the nature of the chert occurrence in the Ocean Floor Assemblage. The chert is a thin part of the rock sequence, indicating a small ocean basin—that is to say, the spreading center was close to the subduction zone. Also suggesting that the ocean ridge magma source was near the arc is interleaving of ocean floor basalts and cherts with arc-derived sands in the Constitution Formation. Following this thought, the Fidalgo ophiolite arc (described in the next chapter) would seem handy and a possible candidate for the abundant greywacke sands in the Ocean Floor Assemblage, but its igneous rocks are too old, ~170Ma.

The search continues for a 150 Ma island arc landmass that was once in the vicinity of the San Juan Islands terranes; perhaps this elusive sediment source existed far afield to the north or south along the continental margin, or elsewhere in the circum-Pacific region.

Fig. 60 Deformed and metamorphosed greywacke of the Lummi Formation. The rocks were heated by burial and were very slowly squeezed, aligning platy minerals to form a planar structure, termed cleavage, and making folds in the sedimentary bedding. 60A Slatey cleavage on North Cone Island, examined by Peter Brown. 60B Detail of cleavage, parallel to pencil, overprinted across sedimentary bedding on Eliza Island. 60C Strongly folded bedding that reflects plastic deformation of solid rock. Swiss army knife for scale. South end of Eliza Is.

# CHAPTER 9
# *FIDALGO OPHIOLITE*

Exposed on Fidalgo Island is a geologically special assemblage of igneous and sedimentary rocks that is an uplifted section of the entire thickness of oceanic earth crust from the surface down to the mantle. This rock suite is known in accretionary zones in many other parts of the world and is

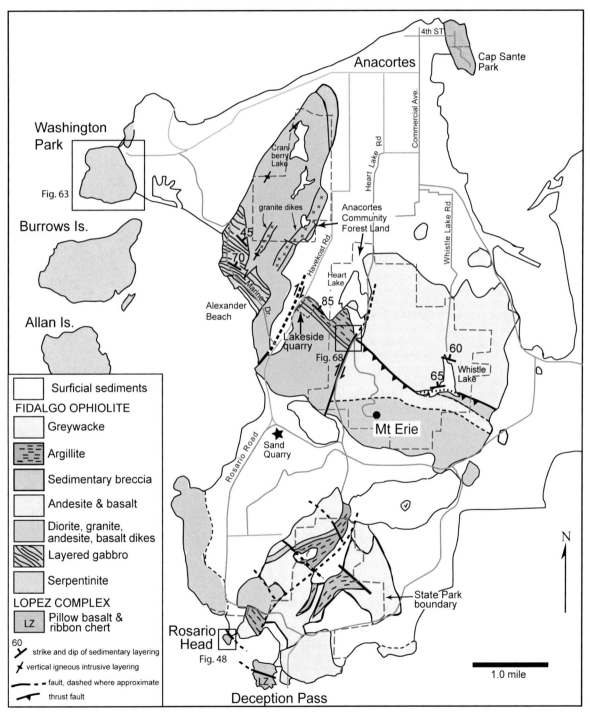

**Fig. 61 Geologic map of Fidalgo Island, adapted from Gusey and Brown (1987).**

termed ophiolite, translated from Greek as green snake stone. Ophiolite crustal sections are on the order of ten kilometers or less in thickness, thinner than continental crust; these rocks formed as part of the ocean crust.

The general order of rocks in ophiolites defines a layered succession from the bottom up of: metamorphic mantle peridotite, igneous peridotite and gabbro, a vertical igneous dike swarm, basaltic lavas, and at the top, ocean floor sediments. Some ophiolites match the geology of a mid-ocean ridge system and are therefore true ocean crust; other ophiolites formed as primitive island arcs. These are two very different tectonic settings. For the Fidalgo ophiolite, an island arc origin is indicated by the make-up of the igneous rocks that are relatively rich in silica, including granites and andesites, found in arcs but not in ridge systems. The Fidalgo ophiolite is best exposed on Fidalgo Island (Fig. 61), but occurs widely elsewhere in the San Juan Islands (Fig. 27): making up most of Cypress Island as peridotite, comprising large parts of Blakely and Lopez Islands as a complex of plutonic and volcanic igneous rocks, and on Decatur and James islands occurring as conglomerate and sandstone apparently derived from erosion of the ophiolite. Alteration, probably by sea water, has changed peridotites to serpentinite, and added greenish minerals to the igneous rocks—thus the "green snake stone".

On Fidalgo Island all the layers of the ophiolite are pretty much intact relative to their formation, so the crustal section can be discerned (Fig. 62). From the base upward are: serpentinite altered from peridotite; layered gabbro; a dike complex of granite, diorite, basalt and andesite that intrudes the underlying serpentinite and the gabbro; lava flows of mainly andesite and basalt; a coarse-grained sedimentary breccia with angular chunks of all underlying rocks; deep ocean clay-rich sediment— argillite; and at the top, a thick section of volcanic-rich siltstone and sandstone derived from an island arc (Figs. 63-66).

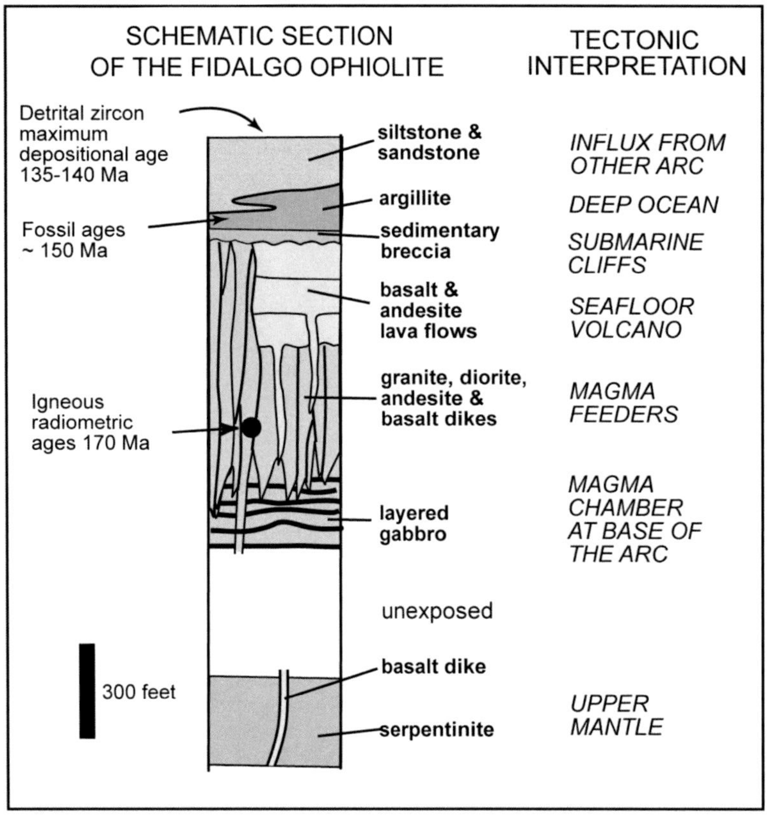

**Fig. 62 The Fidalgo ophiolite stratigraphic section, with inferred oceanic settings.**

Discovery of the Fidalgo Island ophiolite section evolved over a couple of years in the 1970s. Mapping by canoe and on foot, and with the blessing of many property owners, I outlined the distribution of rock types. But the big question was how the rocks were related. A breakthrough finding came from exposures in the Lakeside Industries quarry, off Havekost Road (Fig. 61), where the remarkable relationship is found of deep ocean argillites (mudstone) lying conformably on coarse talus deposits (breccia) and granite. The granitic rocks were apparently intruded and then broken into talus in a marine environment—unusual! A further piece came from mapping the ridges west of Havecost Road as vertical dikes of granite rooted in the layered gabbro along the coast and truncated by the talus/breccia

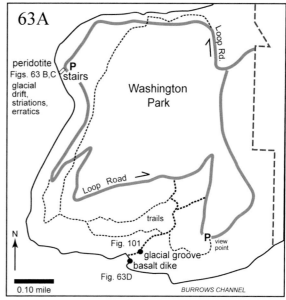

Fig. 63 Bedrock features at the base of the Fidalgo ophiolite in Washington Park, at the northwest corner of Fidalgo Island.

63A Map of Washington Park.

63B Serpentinite, a metamorphic rock derived from peridotite. The peridotite was composed of olivine and pyroxene minerals. In the serpentinite, there are lighter colored orange zones and darker grey zones. The orange zones consist almost entirely of serpentine, altered from olivine; no pyroxene remains. The darker rock has serpentine and abundant pyroxene (in close inspection identified by shiny cleavage surfaces). Basalt magma comes from partial melting of mantle peridotite, and specifically the pyroxene component, leaving the olivine as a residue. Here, we can speculate that the orange zones are residue left from pyroxene melt extraction, and that the darker rock did not give up its pyroxene.

63C Vein of coarse pyroxene crystals crossing the serpentinite. Apparently the vein marks a pathway for pyroxene melt.

63D Basalt dike intrusive into serpentinite. The dike is possibly derived from the peridotite and is likely a feeder to the overlying arc.

Fig. 64 Plutonic rocks in the Alexander Beach area.

64A *left* Light colored granite dikes injected into older gabbro dike rock. Varied intrusions represent pulses from a magma chamber with differing compositions.

64B *below* Graded cumulate bedding in a magma chamber. During cooling, crystals formed and settled to the bottom of the chamber. Layers represent pulses of crystallization, possibly caused by injections from below into the chamber. Dark crystals are mostly pyroxene; light crystal are feldspar. The dark minerals are denser and settle more quickly than the feldspar. The prominent layer graded from dark to light upward in the photo probably formed by crystal settling from a turbid slurry of crystals falling down the flanks and across the floor of the chamber, a sedimentary process akin to that making turbidite sandstones. The layering indicates that the magma chamber has been tilted, and that the direction towards the top of the chamber is upward in the photo.

Fig. 65 Polished slab of the submarine breccia unit from the Lakeside quarry. Rock chunks are all igneous. Coarser textured material is diorite and granite, finer grained rock is andesite and rhyolite. Interstitial material is very fine-grained dark radiolarian ooze. This rock formed as talus along the base of a submarine cliff.

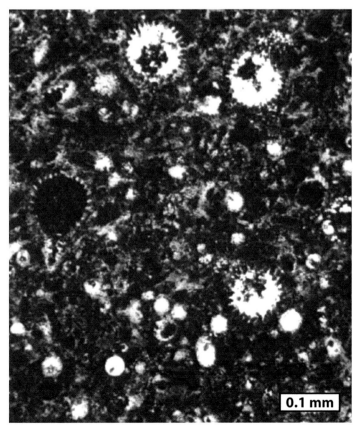

Fig. 66 Pelagic argillite composed of abundant shells of radiolaria, tiny planktonic organisms that accumulated in deep, quiet sea water, away from a landmass shedding sand and silt.

layer north of Cranberry Lake (Fig. 61). There seemed to be a continuous section of rocks from sediments at the top, to granite in the middle, to gabbro at the bottom. Finally, a corroborating observation was that the graded bedding in the cumulate layers of gabbro along Alexander Beach indicates that the top of the magma chamber is to the northeast, toward the sedimentary cap on the igneous rocks near Cranberry Lake.

What type of plate tectonic model can explain the Fidalgo ophiolite?

We start with the evidence that the igneous rocks, of mainly andesite and diorite composition, constitute an island arc, which in turn we think was fed by a subduction zone (Figs. 4, 30). The capping sediments embedded with marine fossils indicate an oceanic setting. In detail, at the base of the ophiolite section we see evidence from a basalt dike intrusive in the peridotite that magma feeding the arc rose through the serpentinite (Fig. 63C) . The layered gabbro represents the floor of the arc magma chamber where differential settling of crystals separated grains into dark and light layers (Fig. 64B). Each couplet of light and dark represents a pulse of fresh magma into the chamber. The residual magma that hadn't yet crystallized was relatively rich in silica and was intruded upward as dikes of granite and andesite (Fig. 64A). The dikes were feeders to an overlying volcanic edifice built up by lava flows. Sedimentary breccia (Fig. 65) overlying these igneous rocks contains fragments of all the underlying rocks and must represent erosion of a cliff caused by faulting cut across the whole ophiolite section (Fig. 67).

All this magmatism and erosion took place in an oceanic setting as indicated by the overlying rock layer of clay-rich sediment (argillite), which contains planktonic radiolarian fossils (Fig. 66) and formed slowly in a quiet deep marine environment. By the time of argillite deposition the arc was dead. Arc activity is dated at about 170 Ma from zircon uranium-lead analysis. A period of quiet ocean floor residence indicated by the argillite occurred at about 150 Ma, dated by the radiolaria. The final event recorded is a great influx of sand and silt derived from a nearby younger volcanic arc. The bulk of the sand grains are in an age range of 140 - 150 Ma based detrital zircon ages dating this younger arc. The youngest population of zircon ages is 136 Ma, providing a maximum age of the sedimentation. A model for this process is illustrated in Fig. 67.

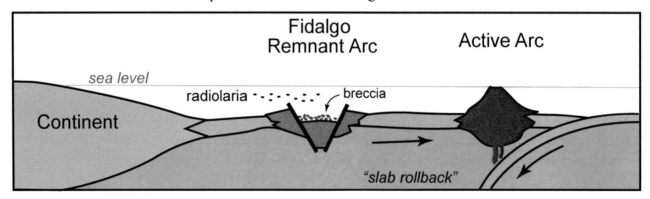

**Fig. 67 Tectonic model for formation of the Fidalgo ophiolite (view south). Initial formation of an island arc volcano, over a subduction zone, produced the igneous rock assemblage of the Fidalgo ophiolite. Then, stepping back of the subducting plate (see also Fig. 30) pulled the subduction zone seaward from the Fidalgo arc, creating a new, outboard active arc. The now extinct Fidalgo arc then lay behind the active arc. It was pulled apart by extensional faulting related to the seaward movement of the active arc, and formed cliffs shedding coarse debris. The "remnant arc" was eventually covered by deep sea radiolarian ooze. Subsequently, sediments from the younger, active volcano covered the Fidalgo arc. Uncertainty remains as to whether the landmass labelled as "Continent" was truly continental, considering that no old zircons are found in the Fidalgo sediments. And perplexing, is that the younger "active arc" that we surmise to be present has not been located in bedrock of the region.**

A modern day analogue occurs in the Philippine Sea at the western margin of the Pacific Ocean (Fig. 5). Here behind the active Mariana arc is the Palua - Kyushu remnant arc, submerged below sea level. The subduction zone once feeding this arc has stepped east to form the Marianas arc and trench, leaving the Palua - Kyushu arc as wreckage in a back-arc basin. Seaward migration of arcs is common, and apparently occurs where the subducting plate is old and dense and literally falls into the mantle, pulling the arc seaward (Fig. 30). Besides the arc and trench migration, this process leads to back-arc spreading and extensional faulting. Geology of the Fidalgo Ophiolite fits well with the modern day tectonism.

Fidalgo Island is well endowed both in fascinating geology and also in much public land where rocks can be observed. Starting with the lowest layer in the section, the peridotite occurs throughout Washington Park and also on nearby Burrows and Allen islands (Fig. 61). An excellent locality is found on the "Loop Drive", about 0.4 miles from the beginning of the loop, where cement stairs lead down to the rocky shoreline (Fig. 63).

Seeing these rocks is a trip to the earth's mantle. The original peridotite has been serpentinized by the hydration of olivine, $Mg_2SiO_4$, to serpentine, $Mg_3Si_2O_5(OH)_4$. The serpentinite is a smooth somewhat orange-colored rock, greenish on fresh surfaces. Some parts of the original peridotite contained pyroxene as well as olivine, and in these rocks the pyroxene is not altered and is observable as crystals with shiny cleavage surfaces a fraction of an inch across. The pyroxene-bearing rock has a darker green-brown look. These two forms of the altered peridotite are intermixed in irregular layers and pods (Fig. 63A). Both types of serpentinized peridotite are crossed by veins of pure pyroxene (Fig. 63B). What are these rocks are telling us about the mantle? The overlying igneous rocks are produced by melting of mantle rocks, and the chief component to yield the right material is pyroxene. Thus we can speculate that the pyroxene-free rocks are depleted of pyroxene by melting; the pyroxene-bearing peridotite has not melted; and the pyroxene veins represent pyroxene-rich melt passing through the mantle.

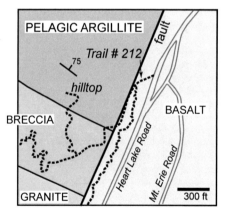

Fig. 68 Locality map for accessible outcrops near Heart Lake in the Anacortes Community Forest Lands.

The important basalt dike intrusive into peridotite (Fig. 63C) is exposed in the coastal area on the south side of Washington Park off the Burrows Channel trail. This dike is likely related to dikes higher in the arc section, providing evidence of linkage of the peridotite to the arc rocks.

Layered gabbros are well exposed (Fig. 64B) along the rocky coast north of Alexander Beach on private property. The rocks can also be seen, but less clearly, in road cuts along Marine Drive. The layering is nearly vertical and in places individual layers are graded from pyroxene rich to feldspar rich upward in the magma chamber—toward the overlying sea-floor strata. Now the whole section is tilted. Granitic dikes seen higher in the arc section are rooted in the gabbro and display sheeted intrusive structure (Fig. 64A).

A locality in the Anacortes community forest land, accessible by trail off the Heart lake Road (Fig. 68), provides outcrops of the arc section including the granitic dike complex, coarse breccia, and overlying pelagic argillite. These exposures are not nearly as good as those once present in the Lakeside quarry.

The topmost unit of the Fidalgo Ophiolite is dark-colored bedded sandstone and siltstone derived from eroded volcanic rocks, and exposed over a large area of the Anacortes community forest northeast of Mount Erie. Access is provided by trails off the Mt Erie Road and Whistle Lake roads.

CHAPTER 10

# NANAIMO GROUP

Rocks of the Late Cretaceous Nanaimo Group record a much less dramatic tectonic history than the old ocean arcs and subducted sea floor formations of the San Juan Islands. These are relatively pristine sandstones, shales and conglomerates derived from erosion of surrounding highlands of Wrangellia, the Coast Plutonic Complex, and the San Juan Islands nappes. By Nanaimo times, these major tectonic elements of the Pacific Northwest were pretty much together, and the continent had grown westward to where we see it now. The Nanaimo Group extends from the San Juan Islands some 150 miles north along the western flank of Georgia Strait underlying the Gulf Islands and the eastern shores of Vancouver Island (Fig. 24).

Fossils date the sediment from about 93 to 65 million years and zircon ages push the origin back to about 95 million years (Fig. 69). The fossils document a dominantly marine setting with sea water depths as great as 5000 feet. Source materials of the sediment are well displayed at conglomerate outcrops where pebbles and cobbles typical of the surrounding hills and mountains are observed - e.g. granite, chert, volcanic rock, argillite, metamorphic rock and limestone.

Aside from some subaerial river deposits along the western margin of the Nanaimo Group, most of the sediment reached its final resting site as turbid submarine flows, indicated by massive bedding, poor sorting, and current-formed bedding structures. An initial on-land origin of the marine sediment is evident from the well rounded shape of pebbles and cobbles pointing to an early history of eroded fragments gaining their rounded forms while bouncing and rolling along in vigorous streams. Thus combining these scenarios, we see that mountain erosion and stream transport created deltas of loose

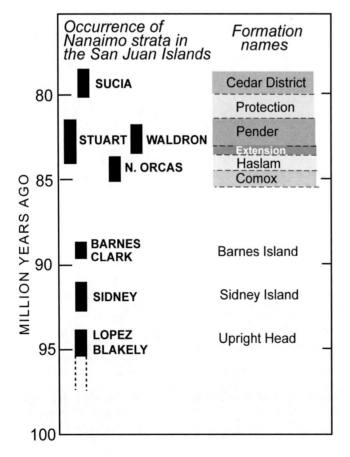

**Fig. 69 Ages and designation of formations of the Nanaimo Group in the San Juan Islands. Sidney Island in British Columbia is included here because it is near the San Juan Islands and is important in constraining ages of thrusting. The older formations, Barnes Island, Sidney Island, and Upright Head, are not as extensively exposed or as well documented as the younger formations. The wide age gaps shown between the older formations may or may not represent gaps in sedimentation.**

58

sediment at the shoreline; these deltas in turn collapsed and fanned out as turbidity flows into the marine basin.

The two and a half mile thick pile of Nanaimo sediments is amenable to subdivision into formations that are distinguished by age and sediment type (Fig. 69, 70). In turn, mapping of these individual formations gives us much information about variation in the sedimentary basin and the structural geometry of the unit as a whole. From this analysis we see the on-land deposits restricted to the eastern flank of Vancouver Island, and a deep marine basin lying between Vancouver Island and the mainland receiving the bulk of the sediment. Mapping of the formations reveals impressive large-scale folding (Fig. 70) that represents coast-perpendicular contraction of the continental margin, dated at about 40 Ma.

Although almost all the material in the Nanaimo rocks comes from the bordering highlands, there is a stranger in the midst. Pebbles and cobbles of pure quartz sandstone (Figs. 21, 71) are found in the conglomerates. Quartz is a mineral concentrated during erosion of granitic (or other) rock by prolonged chemical weathering that has removed less stable minerals. Pure quartz sandstone is not a product of the tectonically active Cretaceous continental margin. It is a common sedimentary rock formed along the edge of the stable passive continental margin, during the early Phanerozoic and Proterozoic eons (350-1400 million years old), now exposed in the Rocky Mountains ~300 miles to the east (Fig. 28). These out-of-place rocks in the Nanaimo group require long distance transport by a river from the Rocky Mountains flowing across the ancestral Cascade mountain belt. A modern day example would be the Fraser River. Some of the quartzite cobbles are half a foot across, indicating strong river currents.

In the San Juan Islands the most extensive exposure of Nanaimo rocks is on Stuart Island, where a section some 3000 feet can be measured (Fig. 70). Other Nanaimo exposures are found along the north edge of Orcas Island (Figs. 72-74), on Waldron Island, the southwest edge of Sucia Island (Fig. 76), and on Clark and Barnes Islands (Fig. 71). These occurrences reveal parts of a

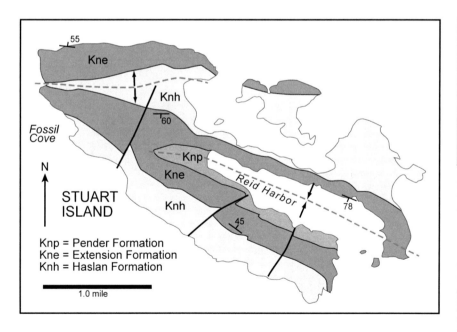
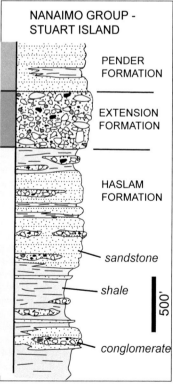

**Fig. 70 Nanaimo Group on Stuart Island. Good coastal exposures and tilting of beds in the large scale folds on the island show the rock details and thickness of the formations. Map from Mercier, 1977. Dashed lines trace fold hinges.**

Fig. 71 Nanaimo rocks on Clark Island.
71A Interbedded conglomerate and sandstone examined by Orion George and Liz Schermer of Western Washington University. This sediment originated by erosion on steep hillsides where chunks of rock were broken loose by mechanical weathering, were then rounded in energetic streams and carried to the seashore. Near-shore deposits were built up and eventually collapsed, perhaps triggered by earthquakes, and swept as debris flows down-slope into the marine basin. The largest boulders are huge (>25 inches). Fossils are plentiful in interbedded siltstones on nearby Barnes Island indicating a marine environment and dating the rock at about 87-90 million years old. An abundance of chert cobbles suggests derivation from the underlying Orcas Chert terrane of the nappe sequence. Volcanic clasts are also common, presumably from underlying nappes.

71B The very smoothly rounded white cobble shown in the photo close-up here, and pointed to by Liz in the photo above, is well-cemented sandstone consisting entirely of rounded quartz grains. This sand required a long slow process of chemical weathering in a setting foreign to the Pacific Northwest environment of steep topography and rapid erosion and deposition. The likely source of this sandstone is in Precambrian strata (~1400 million years old) of the Rocky Mountains, the "miogeocline" of Fig. 28. How the cobble got to the San Juan Islands, by river transport, is a concept to chew on!

sedimentary stratigraphic section in the San Juan Islands that is mostly obscured by seawater and glacial deposits. Clark Island has exceptionally well displayed conglomerates (Fig. 71), with clasts up to boulder size, and relatively easy to spot exotic quartz sandstone cobbles distinguished by their white color and sand grain texture observable with a hand lens. Shale beds of the Nanaimo at the southwest edge of Sucia Island are quite fossiliferous (Fig. 76,77), dating these rocks as about 80 million years old.

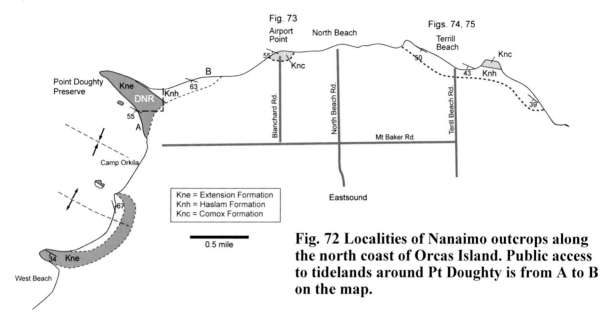

Fig. 72 Localities of Nanaimo outcrops along the north coast of Orcas Island. Public access to tidelands around Pt Doughty is from A to B on the map.

Fig. 73 The Comox Formation at Airport Point, Orcas Island, consists of conglomerate with interbeds of sandstone. Cobbles are chert, volcanic rock and plutonic rock, likely eroded from the underlying East Sound Group and Turtleback Complex.

Fig. 74 Interbedded sandstone and siltstone of the Haslam Formation is well exposed on Terrill Beach, north coast of Orcas Island.

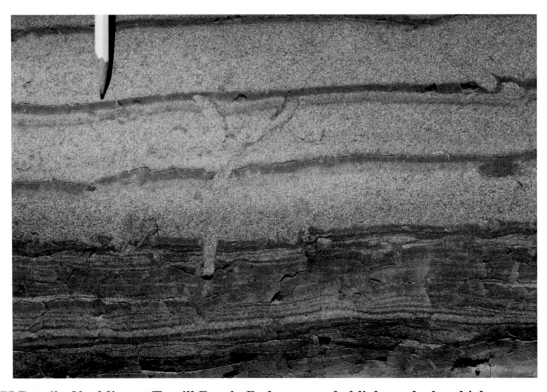

Fig. 75 Detail of bedding at Terrill Beach. Beds are graded light to dark, which represents coarse to fine grain size, going upward in each layer. A slurry of sand and silt sweeping across a marine basin deposited these "turbidites" by settling coarser grains first, then finer as the current slowed. Here, we see a track where a worm ate its way across the soft sediment layers, leaving behind digested sand.

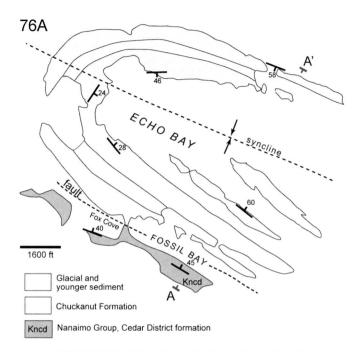
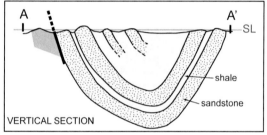

Fig. 76 Sucia Island.

76A Geologic map and cross section of Sucia Island. The Chuckanut Formation and Nanaimo Group are separated by a fault. The arcuate shape of the rock layers at map scale indicates a large-scale fold, in this case a trough-shape that is termed a syncline. The strike and dip symbols in rocks around Echo Bay indicate the angle of inclination of the sedimentary beds at various points around the syncline. Cross section below, along A-A' on the map.

76B Folds in the sedimentary beds, possibly caused by drag along the fault, seen in Fox Cove. George Mustoe photo.

Fig. 77 Inoceramus fossils are ancient clams readily found in the grey siltstone of the Cedar District Formation of the Nanaimo Group along the south side of Fossil Bay on Sucia Island. From the fossils, an age of about 80 million years is known for this part of the Nanaimo Group. The author makes paleontological observations.

A conglomerate recently discovered to be related to the Nanaimo Group occurs at Upright Head on Lopez Island. Outcrops are conveniently accessible on bluffs below the ferry line-up (Figs. 78-80). A similar conglomerate occurs at the north edge of Blakely Island, best accessible by boat. Based on U-Pb zircon ages (Fig. 81), these conglomerates are probably older (~95 Ma) than the oldest Nanaimo fossils (90-93 Ma). They are linked to the Nanaimo Group by the presence of quartz sandstone cobbles, and thus push back the age limit of known Nanaimo deposition.

Fig. 78 Outcrops of conglomerate on Upright Head at the ferry terminal on Lopez Island. The rocks are easily accessed on public land from the parking lot above.

Fig. 79 Abundant pebbles and cobbles of chert and greenstone in the Upright Head conglomerate mostly came from erosion of older nappes in the San Juan Islands. But some light colored pebbles in the photo are quartz sandstone, the same unusual rock of apparent Rocky Mountain source as is found at Clark Island (Fig. 71B) and elsewhere throughout the Nanaimo Group.

Fig. 80 Microscope view shows the rounded edge of a quartz sandstone pebble, and a lumpy internal texture formed by quartz sand grains that are tightly packed and cemented together.

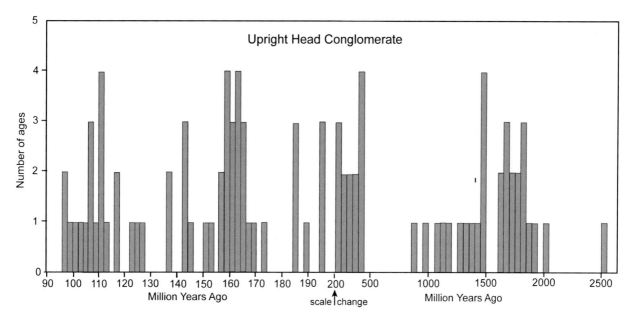

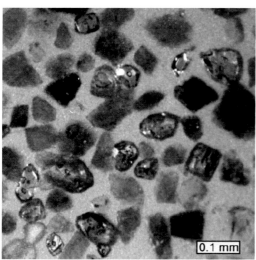

Fig. 81 Ages of zircon grains in conglomerate from Upright Head measured by uranium-lead analysis. The analyzed sample is shown in the microscope photo on the left. Zircons are the bright adamantine grains. The histogram above shows the spread of ages. The old grains, ~1000 million years and older, came from erosion of igneous rocks in the ancient North American continent, Laurentia (Fig 28). Younger grains are from accreted arc rocks along the present continental margin. The maximum depositional age of the Upright Head conglomerate is indicated by the youngest zircon at 97 million years old. The actual age of the rock is likely not much younger considering that younger zircons would have been available from erosion of nearby rocks.

Crustal dynamics during the Nanaimo days were relatively calm, as mentioned above, but not entirely dead. Some evidence points to Nanaimo sedimentation during the latter phases of San Juan Islands thrusting. Zircon in Nanaimo sandstone on Sidney Island, near Sidney B.C., has an age fingerprint of the Turtleback Complex (400 - 500 Ma), indicating that nappe was in place by the ~93 Ma fossil age of this rock. The oldest Nanaimo rock, at Upright Head and Blakely Island (~95 Ma), contains abundant chert cobbles that match with the underlying Orcas Chert unit, indicating presence of that nappe also. But rocks signaling the structurally higher nappe, the Ocean Floor Assemblage, are not present; these rocks show up in the younger Nanaimo strata (~80 Ma) on Sucia and other islands where vein quartz, greenschist, and phyllite are seen (Fig. 82B).

These observations are evidence that the earliest part of the Nanaimo, on Upright Head and Blakely Island, was deposited in an interval between thrust events: after the Turtleback-East Sound and Orcas-Deadman nappes, and before the younger Ocean Floor and Fidalgo nappes. Supporting this idea, the conglomerate on Blakely Island show effects of thrust shearing (Fig. 82A). This interpretation gives us a perception of Nanaimo deposition occurring in an episodically active thrust environment.

Fig. 82   Tale of two conglomerates.

    82A *upper photo* is of conglomerate in the ~80 million year old Cedar District Formation of the Nanaimo group on Sucia Island. Cobbles of metamorphic rock are greenschist, vein quartz, and phyllite that match rocks in the more metamorphosed areas of the Ocean Floor Assemblages, especially parts of this nappe where it extends into the Cascades foothills. Emplacement of the Ocean Floor Assemblage nappe apparently happened prior to the 80 million year age of the conglomerate.

    82B *lower photo* shows conglomerate at the northern tip of Blakely Island, part of the Upright Head Formation, interpreted from ages of zircon grains to be ~ 95 million years old, and based on clast types to have been deposited before the Ocean Floor Assemblage. This rock has been flattened by plastic shear in a fault zone, and therefore apparently was involved in the San Juan Islands thrusting event. The older conglomerate (82B) predates, and the younger conglomerate (82A) post-dates, emplacement of the Ocean Floor Assemblages nappe.

The origin of the Nanaimo basin poses an interesting and unsolved tectonic problem (Fig. 83). Given a basin with a two and a half mile thick pile of sediments that was covered by a mile of water, what happened within this relatively newly formed continental margin to create such a deep (three and a half miles) depression?

The continent at this time was inflicted with folds and faults by the collision of the oceanic Farallon plate. The angle of collision was apparently not head-on, but oblique, and thus the structures record displacements both normal and parallel to the margin. Adding to the mix of crustal instabilities, subduction of the advancing Farallon plate initiated intrusion of massive amounts of magma into the Coast Plutonic Complex, creating a very mobile and active crust just east of the Nanaimo deposits. With these forces in mind, the basin could have been caused by crustal depression related to pile-up of thrust sheets along its eastern margin ("foreland basin"); it could have formed in a pull-apart zone related to margin-parallel strike-slip faulting ("pull-apart basin"); or the basin possibly developed as an extensional sag in front of the Coast Plutonic Complex arc ("forearc basin"). These possibilities, debated in the literature, await further analysis and insight.

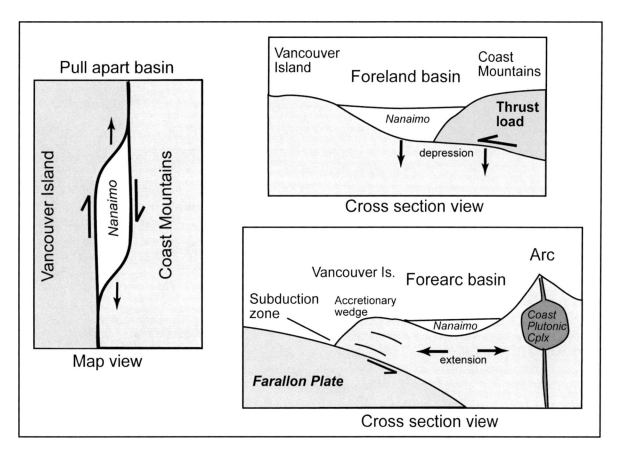

**Fig. 83 Cartoon diagrams of tectonic explanations for the origin of the Nanaimo sedimentary basin. The pull-apart model shows in map view how a steep fault-zone along which Vancouver Island slides north relative to the Coast Mountains (a "strike-slip fault") could open a gap where there is a bend in the fault. In the foreland basin model, crustal depression occurs under a pile-up of thrust sheets coming from the east. The forearc basin explanation shows a convergent margin and a sag in front of the arc due to sinking of the subduction zone in front of the arc.**

# CHAPTER 11
# CHUCKANUT FORMATION

The Chuckanut Formation, made mostly of sandstone and shale, is rich in clues that paint an intriguing vision of a lush coastal landscape in Eocene times. In the San Juan Islands the Chuckanut Formation is the youngest bedrock unit. It comprises the small outer Islands, Patos, Sucia, and Matia, lying at the northern limit of the San Juan Islands and at the south end of Georgia Strait, and it overlies basalt at the north end of Lummi Island (Fig. 27). The major part of this formation occurs on the mainland in the Chuckanut Mountains and in patches across the Cascades (Fig. 24). The Chuckanut Formation developed approximately where there is present-day land in northwest Washington, but the plants, animals and geologic activity of those times were vastly different than now.

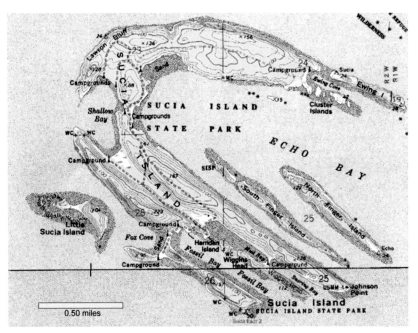

Fig. 84 The topography of Sucia Island is geologically expressive. The elongate Fossil Bay marks a fault boundary between Nanaimo rocks to the southwest and Chuckanut Formation on the main part of Sucia (see the geologic map of Fig. 76). The topographic ridges in the Chuckanut Formation on Sucia and the Finger islands are underlain by thick beds of erosion-resistant sandstone and some conglomerate. The valleys define the intervening beds of softer shale. A curved shape of the bedding, as delineated by topography, represents a large fold plunging gently to the southeast enclosing Echo Bay. In outcrops around Echo Bay the beds dip towards the Bay, thus the fold is a trough—termed a syncline in geologic lingo (Fig. 76).

Chuckanut landscape as we see it now is typically characterized by topographic ridges underlain by thick beds of arkosic sandstone, and intervening valleys cut into shale mostly covered by soil. This is a topography that we see on Sucia Island (Fig. 84), as well as across parts of the mainland.

Features of the sedimentary bedding in the Chuckanut Formation match that of modern river deposits. Cross-bedding in sandstone (Fig. 85) is an indicator of point bar deposits at a bend in the river. Conglomerates are made of gravel and cobbles carried along the bed of river channels. Some conglomerates are near local basement rock and represent deposition by steep side streams (Fig. 86). Mud-derived shales formed in quiet water in overbank floodplains. All these characteristics point to deposition of the sediments in an ancient river system. Fossils of land plants are common in shales, and animal footprints are preserved in mudstone (Figs. 87, 88). Thick coal beds present on the mainland formed from compaction of massive accumulations of swamp vegetation.

Ages of these sedimentary rocks determined from uranium-lead dating of zircon crystals in volcanic layers beneath, within, and overlying the Chuckanut Formation, indicate deposition from about 57 to 45 million years ago, in the Eocene Epoch (Fig. 7).

Fig. 85 Cross bedding in the Chuckanut sandstone along the coast near Larrabee Park on the mainland. The cross beds form in sand bars along river bends.

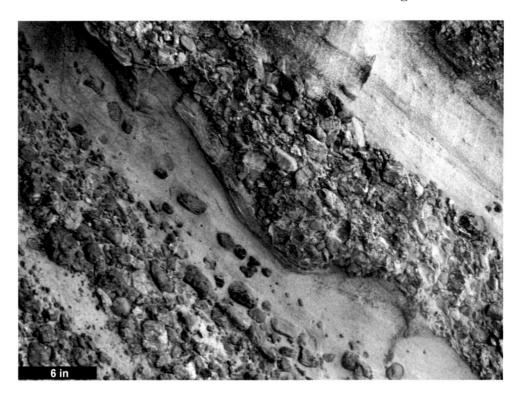

Fig. 86 Conglomerate in the Chuckanut Formation on the coast at the northwest end of Lummi Island. At this locality the Chuckanut Formation lies in depositional contact over the ocean floor rocks of the Lummi Formation. Pebbles and cobbles of chert and basalt in the conglomerate are from the underlying rock. Intervening beds are the usual Chuckanut arkosic sandstone. This outcrop gives us a vision of a landscape of local hills made of San Juan Islands accreted terranes shedding coarse sediment into a broad alluvial plain carrying granite-derived sand from a distant source to the east.

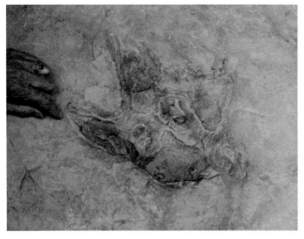

**Fig. 87 Display of features of the Chuckanut Formation, in the Geology Department at Western Washington University.**
*Left photo:* Track of a giant bird in sandstone of the Chuckanut Formation in the Cascade foothills near Mt Baker. This animal is identified as Diatrmya, a flightless bird known from fossils elsewhere on earth to be up to seven feet tall. The absence of marks of talons in the footprint suggests that this bird was not a carnivore, but likely lived off vegetation.
*Right figure:* Artistic illustration of the Diatrmya, with family, in a swampy tropical landscape that represents the origins of the Chuckanut Formation. By Marlin Peterson.

The Chuckanut sandstone is an arkose, rich in quartz and feldspar grains liberated from a granite pluton by just enough weathering and erosion to break away and transport the grains, but not the prolonged weathering conditions that yield quartz sandstone (Fig. 21).

At first thought, the granite source would likely be plutons of the Coast Plutonic Complex in the Cascades. But the mineralogy of the Chuckanut Formation includes a fair bit of potassium feldspar, uncommon in the Cascades granites. A more distant and likely source is granitic terrain in eastern Washington and British Columbia. The Chuckanut Formation and related rocks occur in patches

**Fig. 88 Fossil palm frond in Chuckanut Formation along Mt Baker highway (John Feltman photo).**

extending east across the Cascades, indicating a river system spread across the land that predated the uplift of the Cascade Range. Cross-bedding and other sedimentary structures in the rock show that the rivers flowed generally southwest.

From these observations we envisage that much of the landscape in Chuckanut times was a broad river plain sourced from the northeast. However, some sediment layers are of local derivation. Basalt and chert cobbles observed in conglomerates in the San Juan Islands, for example on Lummi Island (Fig. 86), indicate a source from erosion of the accreted terranes of the San Juan Islands.

Marine sedimentary rocks and fossils the same age as Chuckanut rocks are found in the Seattle area and elsewhere in southwest Washington, indicating that the coastline was not far from the Chuckanut river deposits.

Consistent with Eocene sedimentary rocks elsewhere on Earth, flora and fauna of the Chuckanut Formation, including palm trees (Fig. 88) and tracks made by crocodiles, indicate a tropical climate. Even in polar regions such evidence exists. Relevant to our concern about climate change today, the greenhouse gases of carbon dioxide and methane are estimated for early Eocene times to have been more than double the current atmospheric concentration.

**Fig. 89 Chuckanut Formation sandstone with honeycomb weathering along the shores of Fossil Bay on Sucia Island.**

Adding up the layers of sedimentary rock in the Chuckanut Formation we find, remarkably, a section possibly as much as 5 miles thick! The Chuckanut basin was as deep as the Himalayan Mountains are high. The setting was near the coast where the river deposits were somewhat above, but near, sea-level. During sediment accumulation, the base of the formation sagged to a depth of five miles, but the sedimentary surface was always slightly above sea level.

How this could happen boggles the mind.

We must imagine the sedimentary basin to be continuously subsiding just enough to receive sediment, but not so much as to go below sea-level. Why this balance is maintained is not easily understood. Formation of the basin likely entailed the pull-apart scenario associated with strike-slip faulting that we considered as an origin of the Nanaimo Group (Fig. 83). Supporting the strike-slip idea, the Chuckanut basin was sagging at about the same time the strike-slip Straight Creek-Fraser River fault was moving (Fig. 24) A reasonable plate tectonic explanation is that during Eocene times, plate convergence along the Washington coast was somewhat oblique, as during Nanaimo times, causing a coast-parallel force that led to strike-slip faulting. The sedimentary basin opened where fault strands bent or stepped sideways.

Chuckanut rocks are spectacularly displayed in the Sucia Island State Park, especially in the vicinity of Fossil and Echo bays (Fig. 84). Honeycomb weathering in sandstone (Fig. 89) produces a mosaic of pits separated by thin rock septa, interpreted to be caused by saltwater erosion and hardening of the wall of the pit by algae and other marine organisms. Cavernous tidal exposures in the Chuckanut Formation are sculpted by a combination of wave erosion and surface hardening by chemical weathering (Fig. 90).

**Fig. 90 Cavernous weathering along the north shore of Echo Bay on Sucia Island. The Chuckanut consists of interbedded arkosic sandstone and conglomerate in these outcrops. Bill Harrison and Paul Furlong admiring.**

# CHAPTER 12
# TERRANE TRAVELS

Embedded in rock units of the San Juan Islands are clues of travel histories; the terranes are mostly **not** homegrown. Delving into this history is fascinating in the vision it gives us of crustal mobility on Earth and the processes of building out of North America on its western margin. This aspect of San Juan Islands geology is very much at the edge of certainty—different models are in play; a general consensus is elusive. But for many, this is the fun part. No matter how outrageous, what is possible? How tightly does any model follow the evidence and reasoning?

We start by posing three questions:
- What is the travel history of the individual terranes?
- How were the terranes assembled in the San Juan Islands?
- What displacements moved rocks of the San Juan Islands and neighboring Wrangellia and Coast Plutonic Complex in to their present setting?

We base our answers to these questions by reviewing critical parts of the geology of the terranes that are presented in previous chapters. But also, we introduce a geophysical data set—"paleomagnetism".

Travel history of the terranes
Perhaps the most stunning rock travel issue concerns the history of the island arc comprised of the Turtleback Complex and East Sound Group. As described in Chapter 5, zircons in the Turtleback Complex in the 420-500 Ma range do not match with rocks known in the ancestral western North American continent. They appear to come from a collision zone between Scandinavia and North America along the east side of Laurentia, in part forming the Appalachian Mountains. Travel some 3000 miles to the present locality is presumed to have occurred primarily by back-stepping of the subducting plate which drew the arc along (Figs. 30, 91), moving west on an oceanic plate through the Arctic region to the Pacific realm, and then down the coast to lie outboard of western Laurentia— the "northwest passage". By 360 million years ago the arc was fringing western North America.

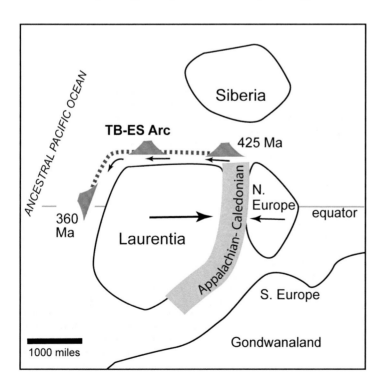

**Fig. 91 Track of the Turtleback-East Sound island arc (TB-ES) in its Silurian-Devonian days. The arc was initiated in a seaway near the Appalachian-Caledonian mountain belt formed by collision of Laurentia and northern Europe as the ancestral Atlantic Ocean closed. The TB-ES arc was active during its path through Arctic to its ending in place outboard of western Laurentia. After arrival near its present locality, the arc carried on volcanic activity until at least 250 Ma.**

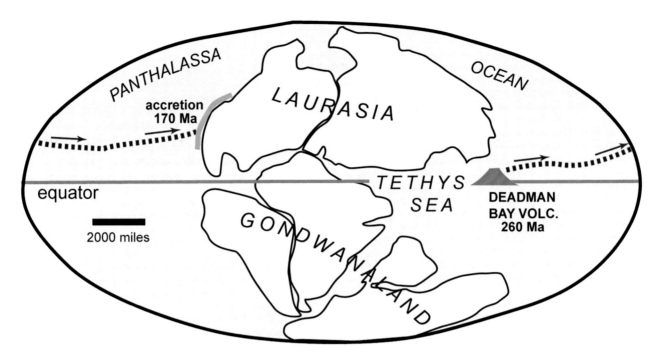

**Fig. 92 The Deadman Bay Volcanics originated in the Tethys Sea. Subsequent ocean plate movement carried these rocks across the broad Panthalassa Ocean to the shores of western Laurentia. Evidence for this distant origin is the occurrence of a particular species of fusulinid that lived only in the Tethys Sea. In this time interval, the continents were all more-or-less together, comprising the "super continent" Pangea.**

Also of distant origin is the Orcas Chert-Deadman Bay Volcanics rock unit, within which fossils (fusulinids) indicate formation in the Tethys sea, at the western edge of the great ancestral Pacific Ocean (Fig. 92). Again, we invoke plate tectonic displacement. This terrane is presumed to have moved 6000 miles or more on a long-lived oceanic plate sliding away from an ocean ridge, towards North America. This possible mechanism is illustrated in Figs. 4 and 47. But, as also for the Turtleback-East Sound arc, we have no remnant marks on the ocean plates to allow a more exact interpretation of how this great travel distance happened. Those old plates were consumed by subduction.

Terranes of the Ocean Floor Assemblage represent ocean crust formed at a ridge system or ocean island. We have very little control over their locations of origin, except to say that they were not much separated from a voluminous source of arc-derived greywacke sand and silt that overlies the oceanic rocks.

The Fidalgo ophiolite island arc is also oceanic in origin, birthplace unknown. But perhaps an indication of proximity to the continent is that this ophiolite is similar to ophiolite in the Coast Range of California (Fig. 28) that has connections to the edge of North America. The Fidalgo ophiolite does not show evidence of high pressure metamorphism (i.e. subduction), which distinguishes its travel history from rocks of the Ocean Floor Assemblage.

How did these terranes become part of North America?

We could surmise that the terranes were swept directly into the continental margin by plate convergence, and in so doing formed the stack of nappes we see now. Such a path is advocated for terrane equivalents of the San Juan Islands running the length of the Cordillera (Fig. 28). But this view of San Juan Islands nappes overlooks evidence for a more complex history indicating that the nappe pile is not an uplifted primary accretionary (subduction) zone: First, the high pressure metamorphism recorded by aragonite crystals and other high pressure minerals is dated to be older than the San Juan Islands thrusts. Thus, subduction followed by uplift happened before thrusting in the San Juan islands. Second, we can deduce that the order of stacking of the nappes is younger upward, based on: higher nappes cross-cutting features of lower nappes, chunks of a lower nappe ripped up into the floor of the upper nappe, and zircon ages that bracket the thrusting age. This stacking sequence of upwardly younger emplacement ages in the nappe pile is the reverse of that seen in accretionary complexes (Figs. 4, 17) such as in the Klamath Mountains where the nappe pile grew by underthrusting.

The history that falls out from these observations is that terranes of the San Juan Islands were accreted and exhumed from a primary accretionary wedge somewhere else, probably at different times and places, and were moved to and assembled in the San Juan Islands over a period of tens of millions of years after initial contact with the continent. The individual travel paths of these terranes pose interesting problems that remain to be solved.

Terrane assembly in the San Juan Islands
We get help in understanding the timing of thrusting in the San Juan Islands from the Nanaimo Group, which was somewhat of a bystander to the process. Sand and cobbles in the Nanaimo strata are in part from the San Juan Islands thrust sheets, and these clasts are, to a degree, distinctive of individual nappes. Knowing the ages of the Nanaimo strata from fossils, and matching sand or cobbles to a particular nappe, we can tell when the nappe was in place shedding sediment into the depositional basin.

Arrival of the nappe made up of the Turtleback Complex and East Sound Group (the basal nappe in the San Juan Islands, Fig. 26) predates the oldest Nanaimo rocks on Vancouver Island, known to be ~93 Ma from paleontology. These Nanaimo rocks contain zircon sand grains distinctive of the

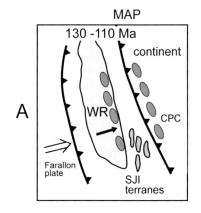
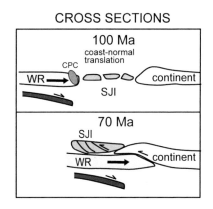
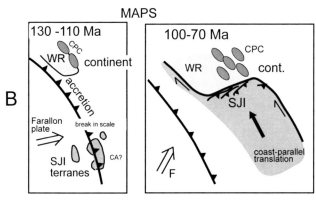

Fig. 93 Two models for emplacement of the San Juan Islands nappes (SJI). Model A proposes that the nappe terranes were moved by plate tectonics into a gap between Wrangellia terrane (WR) and the North American continent. Wrangellia then moved against the continent thrusting the San Juan Islands to the west over the Wrangellian margin. Model B considers that Wrangellia and North America were together and formed a corner area in the continental margin before thrusting. The San Juan Islands terranes moved north along the coast from primary accretionary sites in the south to the corner area where they ramped up over Wrangellia and the Coast Plutonic Complex (CPC).

Turtleback-East Sound nappe. In turn, where this basal nappe of the San Juan Islands extends into the Cascades, it is thrust over a unit (Nooksack Formation, Fig. 25) with zircons as young as 114 Ma. So the beginning of thrusting is bracketed between 93-114 Ma. We can push this minimum age back a bit to be greater than 95 Ma indicated by zircon ages in conglomerate at Upright Head and Blakely Island that contains cobbles apparently from the Turtleback and Orcas nappes (Chapter 10).

Nappe emplacement appears to have been progressively younger going upward in the nappe pile, mentioned above. The youngest nappe emplacement event recognized in the thrust complex involves a sandstone unit in the Cascades with recently discovered zircon grains as young as 74 Ma (Sauer and others, 2014). This sandstone (part of the "Western Mélange" unit) is lumped into a composite nappe that includes the Fidalgo ophiolite.

The overlying and post-thrusting Chuckanut Formation is dated at about 55 Ma in the oldest part. So, nappes arrived and were stacked up in a prolonged event during a period starting before 95 Ma and lasting to at least 74 Ma, and was certainly over by 55 million years ago. This is when the basic architecture of the San Juan Islands was constructed.

How the assembly process occurred is another controversial subject. The plate tectonic setting in these ancient times involved the oceanic "Farallon" plate, now mostly subducted and represented by the Juan de Fuca plate (Fig. 5). In one model, terranes of the San Juan Islands were caught in a collision zone between Wrangellia and the North American margin, squeezed as between the jaws of a vice, in a direction approximately normal to the coastal margin (Fig. 93A). In favor of this idea is the evidence that Wrangellia is of foreign origin, and since it did dock against North America there should be some structural mark of that collision. An alternative model, that I favor, is that Wrangellia was stuck onto the continent before San Juan Islands thrusting, and formed a bump in the coastal margin creating a corner area facing south. San Juan Islands terranes slid one-by-one north along the continental margin until they ran into this corner, and thence stacked up on top of the continental mass of Wrangellia and the Coast Plutonic Complex (Fig. 93B).

But, did the whole assemblage of San Juan Islands -Wrangellia-Coast Plutonic Complex form where it is now?

<u>Displacement of San Juan Islands -Wrangellia - Coast Plutonic Complex</u>
Strong arguments are that this rock grouping came from somewhere south. The "Baja British Columbia" hypothesis proposes the most extreme and controversial model. Geophysicists studying fossil magnetism in rocks use the relation that inclination of force lines in Earth's magnetic field vary systematically with latitude. The inclination is vertical at the north magnetic pole, horizontal at the equator, and varies continuously in between. The needle on a compass feels this effect, and the

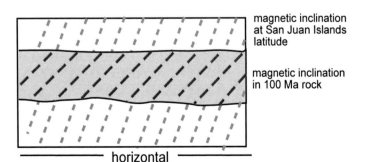

**Fig. 94 Diagrammatic sketch demonstrating paleomagnetic evidence for out-of-place rocks in the Pacific Northwest. Remnant magnetization measured in the terranes corresponds to a much lower inclination of the earth's magnetic flux lines than exists at latitude in the San Juan Islands. This evidence suggests that the rocks crystallized and acquired their magnetization some 2000 miles to the south.**

balance must be adjusted for latitude. Rocks capture the prevalent magnetic field by alignment of tiny iron rich domains (e.g. in magnetite) when they form. Numerous studies of paleomagnetism in rocks of the Pacific Northwest region older than about 50 Ma show that the magnetic field prevalent at the time of rock formation represents lower latitude than at the present position, and equates to an origin of these rocks about 2000 miles farther south than where they are now (Figs. 94, 95). Rocks in the Nanaimo Group as young as 75 Ma record this relatively flat inclination. Younger rocks in the Pacific Northwest (<50 Ma) have magnetic inclination normal for this latitude. In this model (Fig. 95), western British Columbia and northwest Washington formed in Mexico and then cruised north outboard of the coast of California and Oregon at a rate of about 5 inches per year in a period from about 70 to 50 Ma.

The science for this model is hard to refute, but it is not favored by some geologists who ask where is the evidence in California and Oregon for the riding-by of these rocks, and where is the big fault zone in British Columbia along which the moving mass was emplaced. Arguments are made that the magnetic directions measured are not pristine; they were altered by recrystallization, compaction and flattening, or tilting. A further problematic piece of geology is the occurrence of quartz sandstone cobbles in the Nanaimo Group (Figs. 79-81) that have a signature of zircon ages that match Precambrian sedimentary rocks in the northern Rocky Mountains and do not have known correlations to Mexican rocks.

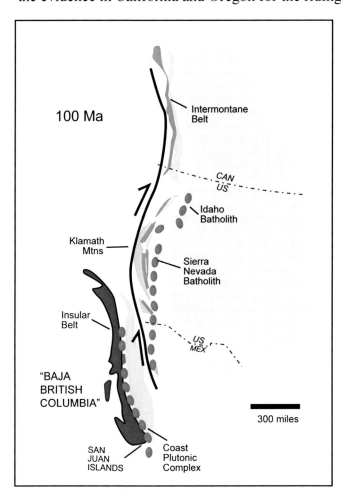

**Fig. 95 Based on paleomagnetism, the San Juan Islands and all of western British Columbia are interpreted to have lain some 2000 miles south of where they are now, for a time period approximately 100 to 70 million years ago. The landmass moved north to its present location by 50 million years ago. This hypothesis is controversial. Compare this diagram with Fig. 28 which shows the present geology.**

A more moderate hypothesis, that pretty much ignores the paleomagnetism, suggests that Pacific Northwest bedrock geology was put together only about 500 hundred miles south, in the vicinity of the Klamath Mountains of southern Oregon and northern California, and moved north in the time-frame of about 100-30 Ma. Direct evidence for northward sliding of this magnitude is the finding of strike-slip faults of appropriate displacement in the Cascades and British Columbia. Also supporting this hypothesis is a match of rock types and ages from the San Juan Islands to the Klamath Mountains (Fig. 28). A modern day analog is the well-known San Andreas fault in California moving Los Angeles at about 2 inches a year up the coast, eventually to pass by San Francisco (after our time).

In conclusion, geology as known today points to tectonic formation of the San Juan Islands by global scale displacements of terranes. Numerous hypotheses are possible for explaining this history, but contradictions and uncertainties remain, fueling the imagination and motivation of many active geologists. A summary of the tectonic history is given in Fig. 96.

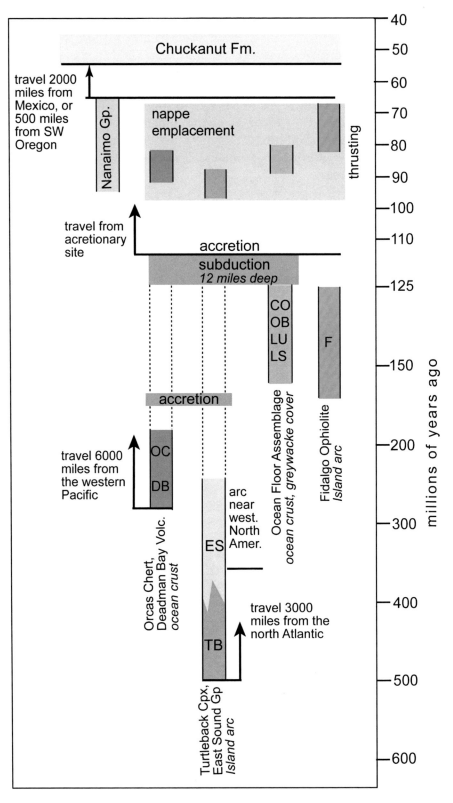

Fig. 96 Time chart of tectonic happenings in the San Juan Islands. The Turtleback-East Sound and Orcas Chert-Deadman Bay terranes formed in far distant locales. We do not have evidence of the accretion of these terranes, but correlative rocks in the Cordillera came ashore about 170 Ma. The Fidalgo terrane may have formed peripheral to the continental margin, suggested by correlation with similar units in California. But the sites of origin of the Ocean Floor terranes are not known. Subduction zone metamorphism is recorded in the Ocean Floor, Orcas-Deadman and Turtleback-East Sound terranes but not the Fidalgo terrane. The locales of subduction are not known. The subducted terranes were uplifted and thence traveled some distance. From roughly 95 to 70 Ma the terranes piled up as the stack of nappes as we see now in the San Juan Islands. Marine sediments of the Nanaimo Group were deposited during thrusting and were somewhat involved in the deformation. All this geology, together with surrounding parts of the Pacific Northwest, was assembled somewhere south along the continental margin and was subsequently displaced northward 100s or 1000s of miles to its present latitude before the Chuckanut Formation was deposited at 50 Ma, marking the end of terrane travels.

CHAPTER 13

# CONTINENTAL GLACIATION

Casual observation of landscapes in the San Juan Islands finds weirdly out-of-place loose rocks on the land surface—rocks that have no relatives in the local bedrock; we call these erratics. A fine example is a huge boulder of granite perched on a point underlain by dark volcanic bedrock on southeastern Jones Island (Fig. 97). These rocks match up with granite in the British Columbia Coast Mountains. They, and a host of other remarkable features of surficial geology, point to glaciation in the San Juan Islands as a part of the global expanse of continental glaciers during the Pleistocene Epoch.

**Fig. 97 Glacial erratic. This granite boulder was carried from British Columbia by the Puget Lobe of the Cordilleran ice sheet and deposited on ancient volcanic bedrock of Jones Island as the ice melted about 16,000 years ago.**

Over the entire period of the Pleistocene Epoch, lasting from about 2.6 million to 12 thousand years ago, many periods of glacial advance and retreat are documented in the higher latitudes of the various continents on earth. The most recent period of global glaciation, termed the Wisconsin Glaciation, lasted from 110-12 thousand years ago. At its peak, about 25-15 Ka, large ice sheets covered most of northern North America, with smaller mountain glaciers blanketing the higher mountains south of the ice sheets. The western part of this great ice mass is known as the Cordilleran ice sheet. It covered all of British Columbia except for the highest peaks and extended south into Washington where it partially covered the Cascades and flowed as a great tongue through the Puget lowland terminating somewhat beyond Olympia (Fig. 98). The glacial geology of the San Juan Islands relates to this last major ice advance.

In contrast to glaciation on most of North America, the lowlands of the Puget Sound region and other glaciated coastal areas of the continent, lay below sea level owing to the great weight of the ice which depressed the earth's crust and uppermost mantle into the plastic asthenosphere. Ice thickness in the San Juan Islands is judged to have been about 4500 - 5000 feet (almost a mile!), as determined from the elevation of ice-carried boulders of British Columbia origin in the Cascade Mountains flanking the Puget Lowland (Figs. 97, 98).

As the Puget Lobe glacier advanced over the San Juan Islands it eroded and sculpted the rock

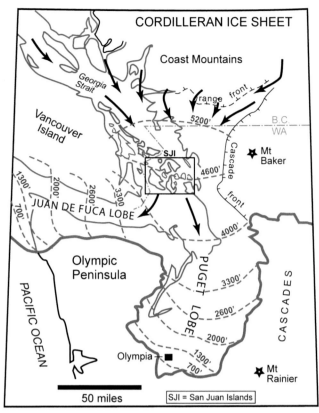

**Fig. 98 Map showing the limits and thickness of the Cordilleran Ice Sheet where it flowed south from British Columbia into northwest Washington. Maximum ice, as shown, occurred at about 17,000 years ago (adapted from Porter and Swanson 1998).**

**Fig 99 View east across the San Juan Islands from the top of Mt Constitution on Orcas Island to Mt Baker in the Cascade Range. The ice sheet lobe covered all this territory up to the Cascades. Glacial erratics from British Columbia are found at the 5000' level on the flank of the Cascades, at about the bottom of the snow in the picture, indicating an ice thickness of close to a mile. Clark and Barnes Islands are in the left foreground; Lummi Island spreads across the center of the picture.**

landscape, evidenced now by grooved, striated, polished, and streamlined bedrock (Figs. 99-102, 104). The glacier carried much rock debris eroded from its floor and flanks, and where the ice melted, either at the front of the ice lobe during glacial advance or over the broad glaciated region when the ice sheet melted, sediments—termed "glacial drift"—accumulated in these places (Figs. 103-106). The glaciated landscape presents a mixture of relatively smooth soil surfaces of drift, and rugged outcrops of eroded rock (Fig. 107). From the glacial drift we learn much about the event. Ages are obtained primarily from wood caught up in sediments below and above glacial deposits (older and younger than the glacier, respectively) and are radiometrically dated by $^{14}C$. Ice advance began across the San Juan Islands at about 18,000 years ago, and had retreated by 16,000 years ago (data of Porter and Swanson, 1998). The Puget Lobe, fed from the great ice fields of British Columbia, advanced and retreated as a consequence of climate change.

During the first few thousand years after the glacial maximum, as the ice retreated, big adjustments occurred in sea and land elevations. At the beginning, the land surface was much depressed due to the weight of the ice. Adding complexity to our visualization of the glacial landscape, global sea level was about 350 feet lower because a vast amount of water was held in the glaciers. As the load of ice melted, both land and sea rose.

Understanding this dynamic process in the Puget Lowland is illuminated by the age and nature of glacial sediments. Virtually all glacial sediments in the San Juan Islands at less than approximately 400' elevation are of marine origin—they contain marine fossils (Fig. 105). Most of these sediments are poorly sorted mixtures of silt, sand, pebbles and cobbles. In some places we find large glacier transported boulders, erratics (Fig. 97). The sediments seem to have been dumped quickly either off the edge of a grounded melting ice sheet, or dropped out of floating ice blocks or rafts into the sea. This type of glacial deposit is termed "glaciomarine drift".

How does the occurrence of the glaciomarine drift presently above sea level as much as 400 feet relate to the much lower sea level of Pleistocene times?

**Fig. 98 Puffin Island, off the southeast side of Matia Island, shows impressive sculpting and grooving from over-riding glacial ice that moved from left to right. View east.**

Fig. 101 Giant glacial groove with striations in Washington Park, Fidalgo Is. View is toward Burrows Channel.

Fig. 102 Glacial polish and grooves on the surface of pillow basalt at Richardson, Lopez Is. The high polish is testimony to very fine abrasive sediment, "glacial flower", embedded in the ice.

Fig. 103 Massive deposit of glacial outwash sand lies over bedrock at Cattle Point, San Juan Island. This is a marine deposit from melt water outflow at the margin of a glacier grounded on bedrock (see Fig. 104 below).

Fig. 104 Grooved and striated bedrock, erratics, and poorly sorted glaciomarine drift at Cattle Point, below the expanse of outwash sand in the above photo. The bedrock was scoured during glacial advance and covered by glacial drift during glacial retreat. Linda Brown enjoying the view.

Fig. 105 Glaciomarine drift on Jones Island. A poorly sorted deposit of pebbles, sand and silt contains clam shells that were incorporated during sedimentation in sea water. The large shell is about 2 inches across.

Fig. 106 A thick deposit of glacial outwash sands on Fidalgo Island is exposed in a quarry near the intersection of Marine Drive and Rosario Road. Bill Osborne in the distance.

Fig. 107. Lidar image of Orcas Island. Exquisite details of topography are revealed by this pulse-laser based photographic technique that records the precise the elevation of the land surface without the masking effect of trees. Here, we see lumpy topography where bare rock is exposed. Linear crevices reflect fractures and faults in the bedrock. The very smooth areas are regions of soil that is mostly glacial drift. Grooves in the drift document the flow directions of the ice sheet: the ice flowed southwest across the island. The entire island was covered by the ice sheet. While some areas were scoured and eroded, other areas received sediment. Study of this image together with the geologic map below (from Fig. 27) further helps our understanding of the landscape.

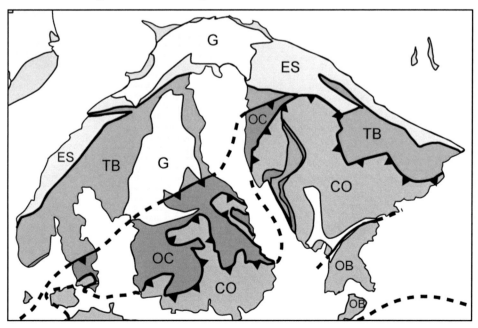

The lidar technique involves bouncing a laser beam from an aircraft off the ground surface. The position of the aircraft is very accurately known from GPS location. The return time of the laser beam is a function of distance. Thus knowing aircraft position and the speed of light, the ground elevation can be calculated. This image is from the Puget Sound Lidar Consortium.

Global sea level during continental glaciation, measured by old shorelines on lands where there was no glaciation or vertical structural displacements to raise or lower local sea level, was approximately 350 feet lower than now. With the knowledge that glaciomarine drift is found up to 400 feet above the present sea level and that sea level has risen 350 feet globally due to melting ice (termed eustatic), it is apparent that about 750 feet of crustal uplift (termed isostatic) has occurred in the San Juan Islands due to rebound from de-glaciation (Fig. 108). This rebound relates to the floating equilibrium of the earth's crust on the plastic asthenosphere (Chapter 1), comparable to a block of wood in water—when a load on top is released the block floats higher. Estimates are that at the start, isostatic uplift was pretty quick in the Puget Sound area—perhaps a half a foot a year. That would be noticeable at beachfront property!

What caused the global climate change leading to continental glaciation of the Pleistocene Epoch?

As measured by the oxygen isotope composition of trapped air in ice cores from Greenland and Antarctic glaciers, climate in these polar regions vacillated by 15-20 degrees Fahrenheit from glacial to interglacial periods. Many heating and cooling cycles are recorded in the ice cores: five in the last 450 thousand years. These data indicate cool periods of roughly 100,000 years separated by 10,000 - 20,000 year warm periods. The periodicity of the temperature cycles matches well with certain episodic variations in the earth's orbit and tilt ("Milankovitch cycles"), providing convincing evidence for an astrological cause. However, there must be more factors because glacial and interglacial cycles have not occurred throughout geologic time.

We know that patterns of circulation of ocean water also influence climate, for example the "El Niño" and "La Niña" events of the modern day. Ocean currents are subject to control by the position of land masses, which are in turn moved about by plate tectonics. But these tectonic displacements are slow, and the Milankovitch cycle, being an ongoing climate force, should prevail for a long time. Barring human impact on climate, the Juan de Fuca lobe will be back!

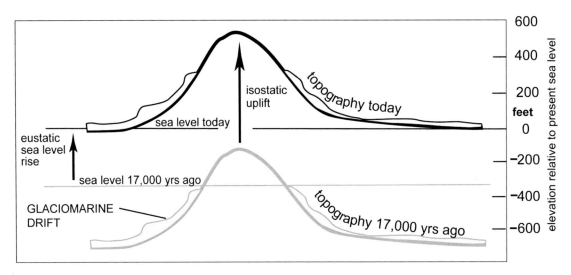

**Fig. 108 Diagrammatic illustration of the effects of global sea level rise and local isostatic rebound in the San Juan Islands since the retreat of the Juan de Fuca and Puget lobes of the Cordilleran continental glacier. Soon after the ice sheet began wasting at about 17,000 years ago, sea water moved in over the lower elevations of San Juan Islands. Much ice-carried sediment was distributed in the marine waters as "glaciomarine drift", now exposed above sea level owing to the crustal rebound, isostatic uplift. Global sea level rise due to melting of continental glaciers, eustatic change, was about 350 feet. Total isostatic uplift due to crustal rebound as the weight of ice was removed was about 750 feet.**

# PART III – FIELD GUIDE

## CHAPTER 14
## *GOOD PLACES TO SEE GEOLOGY IN THE SAN JUAN ISLANDS*

This book is organized primarily to tell the story of how the San Juan Islands came to be, with focus especially on the tectonic evolution. With this purpose in mind, the chapters are organized to describe and interpret the geologic features in the order of their formation.

This last chapter serves as a more systematic guide to each island for seeing the geology. Some additional location maps are presented here, but much reference is made to materials given in the preceding chapters. Otherwise, the general map (Fig. 109) together with directions and a roadmap should be sufficient for finding the locales. Some places are accessible only by land, others only by

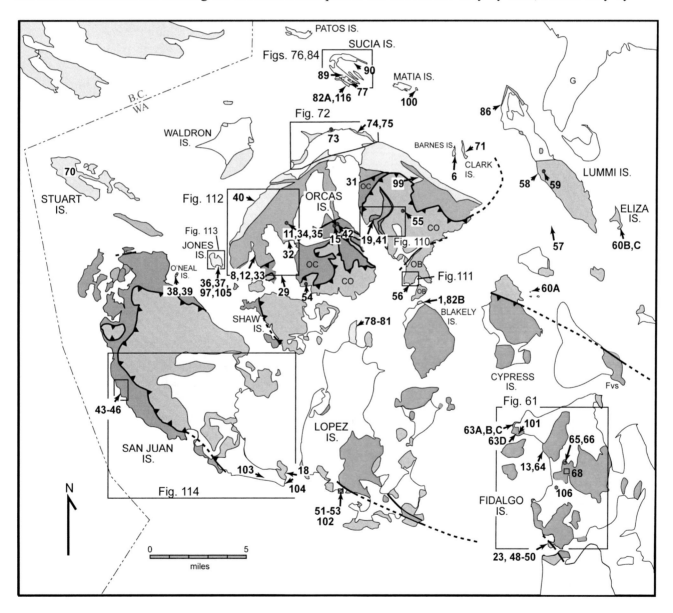

**Fig. 109** Map showing localities of maps and photos occurring throughout the book. The legend for this map is given in Fig. 27.

water, and many either way. The sites listed are relevant to the prevailing goal of the book, and are places I have studied; but there is much other terrain in the Islands worth a look.

As a general rule, outcrops in the upper level of the tidal zone provide by far the best exposures in the Islands. The rocks are clear of soil, vegetation, and stain, and typically reveal details down to a level worthy of examination with a hand lens. With this in mind it is useful to know what coastal areas are accessible to the public. The Washington State Coastal Atlas provides this information: https://fortress.wa.gov/ecy/coastalatlas/tools/Map.aspx. Thus, besides the localities described here, virtually any place you find yourself at shoreline will have geology of interest.

## Orcas Island

A must-visit locality is the top of *Mount Constitution*. To get there, find your way to Moran State Park (Fig 110), and thence up the steep and twisty paved Mount Constitution Road to the top. About half-way up at one of the many hair-pin bends, you come upon a pull-off with a fabulous view south. Also here is a chance to see outcrops of the Constitution Formation (such as they are) on the uphill side of the road a few hundred feet back down the road from the viewpoint (zircon ages, Fig. 55). Drive on to the parking area at the top, where you will need a state park permit. From here it is a short uphill walk (~200') to the summit— 2409 feet above the sea. The views are to the north, east, and south looking down on other islands and into the Cascades (Fig. 99). The stone tower here, built by the Civilian Conservation Core in 1936, gives an even broader vantage point.

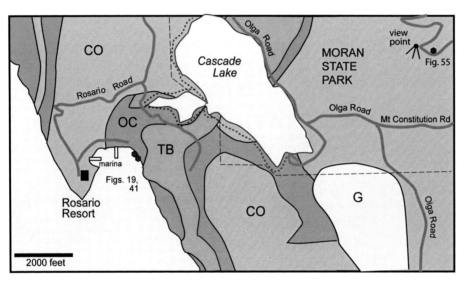

**Fig. 110 Detail of geology in the vicinity of Moran State Park and the Rosario Resort.**

A visit to the *Rosario Resort* (Fig. 110) also provides geologic opportunity. From the Olga Road, head down the Rosario Resort Road, and as you get to the main hotel take a left past the marina where you can park near the waterfront. The tidal zone is open to the public here and gives access to excellent outcrops of the Orcas Chert, and some fault zones (Figs.19, 41).

*Obstruction Pass State Park,* a next good destination, is down the road to the south. Go on Olga Rd, to Point Lawrence Rd, to Obstruction Pass Rd, to Trailhead Rd., and into the north edge of the Park (Fig. 111). Trails in

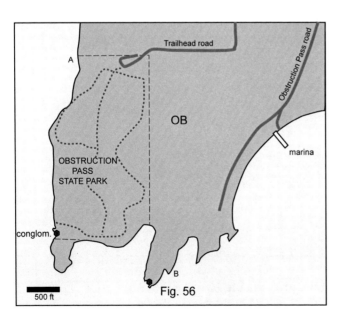

**Fig. 111 Map of Obstruction Pass State Park and surroundings.**

the Park take you to the beach about half a mile to the south. In addition to the public lands of the Park, the tidal zone is public access around the Park and beyond, from A to B on the map. The Obstruction Formation here is mostly well bedded sandstone and siltstone (Fig. 56). An interesting conglomerate at the southwest corner of the park, has clasts of chert, limestone, sandstone and blocks of black argillite up to a foot across. Some geologic history remains to be figured out here.

Good geology stops abound in *southwest Orcas Island* (Fig. 112). The trails in the Turtleback Mountain Preserve wander through a large expanse of rocks of the Turtleback Complex and the East Sound Group. These rocks are not as well exposed here as in coastal areas, but the side benefits of a good hike and great views make this area highly worthwhile. Check the website for the San Juan County Land Bank for more information.

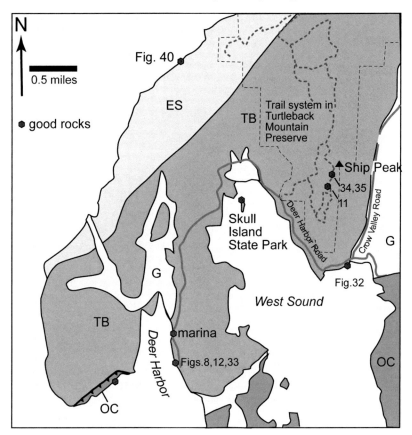

**Fig. 112 Geologic map of southwest Orcas Island with sites of geologic interest.**

Easy access and good exposure of Turtleback rocks are seen along the beach at the Deer Harbor Marina, and a few hundred feet south along the coast (Figs. 8,12,33, note that this locality is not designated as public access).

Some other good geology can be found by water access on public land or public tidal zone (Fig. 112): Turtleback granites at Skull Island State Park, ribbon cherts in the Orcas unit on the west side of Deer Harbor, and the best greywacke turbidites in the islands along the southwest coast (Fig. 40).

Along the *north coast of Orcas Island* rocks of the Nanaimo Formation are well exposed (Fig. 72). Public access by land is unfortunately limited to one locality, at Airport Point (Fig. 73). Travelling by water, a good stop for examining conglomerate of the Extension Formation of the Nanaimo Group is at Point Doughty. Besides its geological attraction, this DNR land is a popular seaside camp for kayakers. The extent of public tidal access lies between A and B on Fig. 72.

Finally on Orcas Island, while you have plenty of spare time in the ferry lineup, check-out the road cuts of Constitution Formation and see the green color due to fine metamorphic minerals and angular fragments of volcanic rock (Fig. 54).

## Lopez Island

Lopez Island has many charms: much undeveloped rural landscape, a neighborly custom of waving to passers-by, and many public access points to the coast; it is my favorite of the Islands. In terms of geology, two localities stand out: Upright Head and Richardson. The rock of *Upright Head* is mostly conglomerate. This rock can be well examined on public access land on the east side of the Head at the ferry terminal area, and by the dock at the north end of *Odlin County Park*, a mile south of the ferry terminal. At both localities you can find the exotic quartz sandstone cobbles, apparently carried from the ancestral Rocky Mountains (Figs. 78-81). Also see the very abundant chert cobbles and granitic rock from the Orcas Chert and Turtleback Complex, respectively. This conglomerate is apparently the oldest formation of the Nanaimo Group.

The *Richardson* locality is at the south end of Lopez Island, about nine miles from the ferry. Go south on Ferry Rd, east on Fisherman Bay Rd, south on Center Rd, west on Lopez Sound Rd, and south on Richardson Rd to the end. Rocks here record intense tectonic activity in Cretaceous time, described in Chapter 8 (Figs. 51-53), and a much later story of glaciation, Chapter 13 (Fig. 102).

## Jones Island

The entire Jones Island is a lovely state park, complete with dock, buoys, campsites, fresh water, and excellent coastal rock exposures accessed by boat or on foot (Fig. 113). Bedrock of the island is mainly volcanic rock of the East Sound Group. Explosive fragmental rock (breccia) of andesitic composition is prevalent (Fig. 36). Massive rhyolite is also found; it is light colored and dense. The explosive nature and andesite to rhyolite composition of the igneous rocks is a classic signature of arc volcanoes, as opposed to the basalt lava flows of ocean island or ocean ridge volcanic settings. Sedimentary rocks are represented by greywacke turbidites and limestone. The limestone featured in Fig. 37 is particularly intriguing in that it is interlayered with sandstone and intruded by dark igneous rock that is andesite or basalt. Probably we are seeing features of a shellfish reef on an active volcano.

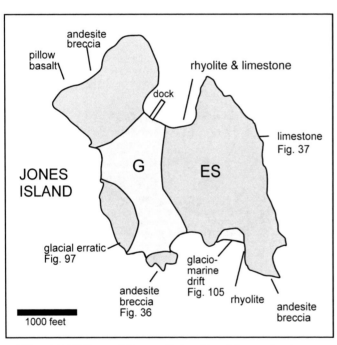

**Fig. 113 Map of Jones Island, with geologic notations.**

## San Juan Island

My favorite geologic and scenic spots on San Juan Island are at Limekiln Point, American Camp, and Cattle Point (Fig. 114). *Lime Kiln Point*, described in Chapter 7 (Figs 44-46), attracts both whale watchers and geologists. The latter marvel at the occurrence of ocean island rocks with fossil evidence of plate tectonic transport many thousands of miles across the ancestral Pacific Ocean (Fig. 92). I have looked for, but never seen the fossils; they are reported to be whitish, about the size and shape of a pinto bean, and tend to blend into the limestone host. There is much to see here: pillow basalts, limestones inter-grown in the basalts, the historic lime kiln itself, large limestone quarries, birds, and whales.

In the *Cattle Point* vicinity is a broad tract of public land accessible to trail hiking and shoreline browsing. A good geology traverse starts at the Cattle Point parking area (Fig 114). The tops of outcrops in this vicinity are grooved and scoured by over-riding ice. Bedrock structure is best studied at the shore, down a somewhat tricky path when the tide is not high. This rock is part of the Lopez Structural Zone, so named because the sandstone/siltstone rocks are much deformed (Fig. 18). Once at beach level, and having examined bedrock structures at the base of the trail, walk out to Cattle Point (when the tide permits). Here you find a continuation of deformed sandstones in the bedrock. Here also the evidence of glaciation is impressive: glacial grooves, striations, and overlying poorly sorted glacial deposits (Fig. 104). The ice-bedrock contact here was hundreds of feet below sea level. A steep trail heads up through a thick section of glacial outwash sands (again, all deposited below sea level) to the plateau above where the Cattle Point Lighthouse is located.

A visit to *American Camp National Historic Park* finds many attractions, historic and natural. I've had my lunch here at South Beach numerous times, sitting on the gravel, back against a log, enjoying views of the Straits of Juan de Fuca and the Olympic Mountains. Orca whales commonly traverse this coast, feeding in upwelling currents. Outcrops of the Orcas Chert unit run along the coast starting at the northwest end of the beach. The Orcas Chert rock here is somewhat famous among geologists for the small-scale markings of deformation that might allow interpretation of the big picture of how the San Juan Island nappes were emplaced. Shear zones, striations, and folds were made during one or more of the many dislocations these rocks have experienced. Some interpret that the nappes slid in towards the northwest, others say to the southwest. Wander along in this complex geology and find your own best interpretation. Trails along this coast lead up the hill past the historic military camp and to the visitor center (also reached by road access from above).

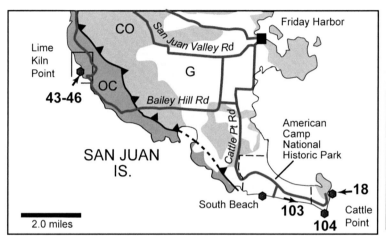
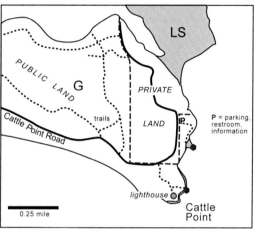

**Fig. 114 Geologic sites on southern San Juan Island.**

## Sucia, Matia, Patos, and Clark Islands

These small islands lie along the northeast fringe of the San Juan Islands. All are State Park lands, with camping, trails, and various levels of boat anchorage. Patos and Matia are entirely underlain by the Chuckanut Formation. Clark Island is made of conglomerate (Fig. 71) that looks like Chuckanut Formation, and was originally identified as such. But fossils in nearby related rock of Barnes Island (privately owned) are of an age (~88 Ma) that indicates affinity to the Nanaimo Formation.

*Sucia Island* has rocks of both the Chuckanut Formation and Nanaimo Group, separated by a fault (Figs. 76, 115). A good start to enjoying the natural wonders on Sucia Island would be to visit the **kiosk** located on a knob along the trail between the two docks (Fig. 115). George Mustoe of Western Washington University created the panels, one of which is devoted to geology. Heading out to look at rocks, check out the shores of Fossil Bay. On the northeast side is Chuckanut Formation, exhibiting tan arkosic sandstone, cross-bedding, and honeycomb weathering (Fig. 89).

**Fig. 115 Detail of the Fossil Bay area of Sucia Island.**

Moving around the head of the bay, and into the shores of Fox Cove, folds can be seen in the Nanaimo rocks at the beach level. This outcrop is on the trace of the fault that has apparently placed the Chuckanut and Nanaimo rocks in contact (Figs. 76B, 115). On the bank above the folds, the Chuckanut Formation is found. The way the folds are oriented, we could interpret that the Nanaimo bedding was bent as the Chuckanut Formation rode over, along a thrust fault (all George Mustoe observations).

On the southwest side of Fossil Bay, across the fault, is grey marine siltstone of the Nanaimo Group. Walking along here you can see fossil clams (Inoceramus), 80 million years old (Fig. 77). Continue southeast around the point, going down-section across bedding into older rocks, and see a marked change in the sedimentary layers—sandstone and conglomerate are predominant here. The combination of wave sculpturing and distribution of bright white quartz cobbles and pebbles in these bedded rocks is a sight to see (Fig. 116).

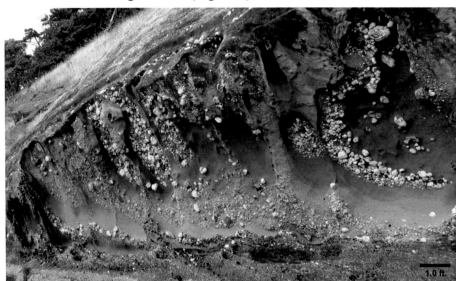

**Fig. 116 Wave sculpted outcrop of conglomerate interbedded with sandstone in the Nanaimo Group, Sucia Island. White cobbles are vein quartz. The outcrop is best accessed by water.**

Of great interest to understanding when San Juan Island nappes were emplaced, the conglomerates around the point contain abundant pebbles and cobbles apparently derived from erosion of Ocean Floor Assemblage nappe and related Cascades rocks. These rock types include the white vein quartz, mentioned above, dark grey slate and phyllite, and greenschist (Figs. 82A, 116). From these outcrops we find that the Ocean Floor Assemblage nappe had arrived in the San Juan Islands by 80 Ma.

## Fidalgo Island

Fidalgo Island is blessed with abundant public land: Washington and Cap Sante parks in Anacortes City, the Anacortes Community Forest Lands, and the Deception Pass State Park—all with trails and good access to interesting geology (Figs. 61-68).

The knob that you drive to in *Cap Sante Park* offers, besides excellent views, good evidence of glaciation in the rounded, grooved and striated bedrock surface. The rock itself is a coarse sandstone, rich in chert fragments, that is deformed and metamorphosed. Better exposures of this bedrock for examining textures are seen along the shoreline, where tidelands accessible to the public are conveniently reached at the east end of 5th street. Here, the clean outcrops show flattened grains defining a metamorphic foliation, and cross-cutting quartz veins. This rock is certainly more deformed and metamorphosed than sandstone in the Fidalgo ophiolite which makes up most of Fidalgo Island. The Cap Sante rock has affinity to the distant Lummi Formation, and has somehow been faulted into place here.

*Washington Park* is entirely underlain by peridotite, the olivine- pyroxene rock that is an uplifted fragment of the earth's mantle and constitutes the basement of the Fidalgo ophiolite. These rocks are easily accessible off the Loop Road, down steps onto the shoreline (Fig. 63). Additionally here, at the bottom of the steps, is a flat glacier-beveled rock surface with striations, unsorted deposits of glacial drift, and glacially carried granite boulders from British Columbia.

Another impressive geologic locality in Washington Park is reached by a short hike off the Loop Road, about one and a quarter miles around, down either of two trails shown by heavier pattern on the Fig. 63A map. The trails take you about 1/4 mile down to the coast at Burrows Channel. Almost to the bottom, on the left, is a huge glacial groove (Fig. 101). At the water's edge, somewhat around the corner to the left as you are facing the water, is a thick basalt dike intrusive into the peridotite (Fig. 63C). It is somewhat tricky climbing around here on slippery rocks.

Moving on to inland parts of Fidalgo Island in the *Anacortes Community Forest,* hike trails into Whistle Lake to see good outcrops of the greywacke rocks at the top of the Fidalgo Ophiolite section. Come in from the north off the Whistle Lake Road or from the Mt Erie Road (Fig. 61).

At another locality, hike a trail system off the Heart Lake Road to observe several geologic units, albeit very marginally exposed (Fig. 68). As you start up trail 212, across the highway from the parking place at the bottom of the Mt Erie road, you walk over dark angular chunks of basalt coming up through the soil from bedrock below. This is the volcanic section of the ophiolite (Fig. 62). A few hundred feet along, the basalt chunks are no longer seen, and you get to an outcrop of the sedimentary breccia which has angular fragments of granite and basalt (discerned with difficulty here, but more obvious farther along). This deposit formed as a submarine talus pile off cliffs of the Fidalgo ophiolite. On the trail between the basalt chunks and breccia outcrop, you passed over a big strike-slip fault in the bedrock. Farther on, about a quarter of a mile from the highway, a trail comes in from the right (northeast). As you go up this trail-branch to the hilltop, the pelagic argillite is revealed in small angular chips in the soil, and reclusive outcrops. This rock is very fine-grained, black, and hard (Fig. 66).

The summit of Mt Erie is a must-do. On the 1.5 mile trip to the top (el. 1273') you pass dark outcrops of basalt, the same rock seen as chunks along trail 212 described above. At the top, the view is supreme. The Cascade range is seen on the horizon to the east, where the glacier-clad Mt Baker volcano is a dominant feature. The Twin Sisters range, rusty-colored peaks just to the south of Mt Baker, are underlain by a six-mile-long horizontal slab of peridotite—somehow thrust up from the earth's mantle. These and other Cascades peaks stood above the continental glacier lying in the Puget lowland. To the south, the view is down the path of the Puget Lobe of the continental glacier. The low-profile islands are made of glacial deposits. Looking west we see the Olympic Mountains. This range is made of a stack of oceanic lava flows pushed on land beginning about 50 million years ago, and a younger uplifted accretionary wedge of sedimentary rocks related to subduction of the Juan de Fuca plate.

Mount Erie was, of course, over-ridden by the glacier. Along the top of the mountain, especially near the parking lot, you see the streamlined, grooved, and striated bedrock. By contrast, at the edge of bluff to the south, the downstream side of the mountain was quarried and steepened as the ice moved across and plucked off large rock chunks. These cliffs are a destination for rock climbers and hang-gliders. Underfoot, the rocks are mostly stained, but some good exposures on the bluffs show the mineralogy and texture of diorite, and some lighter and darker colored intrusive dikes.

At the south end of Fidalgo Island, the *Deception Pass State Park* (Fig. 61) gives broad access to rocks of both the Fidalgo Ophiolite, and faulted-in parts of the Ocean Floor Assemblage. There is much to see here, and as usual, considerable structural complexity. My favorite is a coherent ocean crustal section well exposed at Rosario Head (Figs. 48-50).

##### ***** FINAL THOUGHT *****

Every human's view of the world is shaped by what they know. Travelers in the San Juan Islands are immersed in nature, and their lives are enhanced by this experience and knowledge gained. Compared to whales, herons, madronas, and all other alive and active characters in the mix of nature in the Islands, geology is less upfront because it is relatively static. However, the geologist looking across the San Juan Islands landscape of erratics, radiolaria, and basalt pillows, has a vision of action—big action! We live in a moment between much that has happened, and will happen. I hope this guide to San Juan Islands geology helps move the traveler out of the moment into a broad dimension of time and change in the wonder of nature.

# GLOSSARY

**accretion.** Adding of rock masses to the continental margin.

**accretionary wedge.** Thick wedge-shaped zone of accreted material at the continental margin, where rock materials on a converging oceanic plate are scraped off and added to the continent.

**ammonite.** Fossil with a distinctive spiral-shaped shell, especially useful in the San Juan Islands for establishing the age of rocks of the Nanaimo Group

**amphibole.** Dark, elongate mineral common in coarse-grained igneous rocks; rich in calcium, iron, magnesium and silicon.

**andesite.** Type of volcanic rock, fine-grained and dark grey, typically with light colored feldspar crystals; common rock in arc volcanoes.

**aragonite.** Form of calcium carbonate, $CaCO_3$. The usual form is calcite, which is the common mineral of limestone. With great depth of burial of limestone (>12 miles), calcite changes to aragonite due to the increased confining pressure

**arc.** Belt of igneous intrusions and volcanoes formed above a subduction zone where the earth's crust dives down and releases fluids at a depth of 50-70 miles that rise and cause melting in the over-riding plate.

**arkose.** Sandstone made largely of quartz and feldspar grains. This sand typically comes from erosion of granite.

**argillite.** Sedimentary rock made of mud and silt; harder and more massive than shale.

**asthenosphere.** Deep layer in the earth, generally twenty to fifty miles below the earth's surface, distinguished by slow velocity of earthquake waves. This zone is close to melting, is plastic. More rigid earth layers above slide across the asthenosphere, allowing plate tectonic transport. When heavy loads are placed on the earth's crust, such as glaciers or sheets of faulted rock, the asthenosphere sinks under the load, providing a type of floating equilibrium.

**atomic structure.** Distinctive latticework of structure in a mineral built up of atoms of the constituent elements. In the case of $CaCO_3$, the atoms of calcium, carbon and oxygen can be organized to form either the mineral calcite, or alternatively aragonite. The two mineral have the same chemistry, but different atomic structure; the aragonite crystal lattice is more compact, reflecting the high pressure conditions under which it formed.

**basalt.** Dark, fine-grained volcanic rock, very common. Basalt magma mostly comes from partial melting of mantle rocks (peridotite) ten to fifty miles deep in the earth.

**Benioff zone.** Spatial pattern of earthquakes where an oceanic plate runs into another plate, typically the continental margin. The starting points (foci) of these earthquakes lie roughly in a plane that dips away from the oceanic plate, under the continental plate, and mark the array of faults caused by subduction of the oceanic plate.

**benthic.** Sea floor habitat of fusulinids in limestones associated with the Deadman Bay Volcanics unit at Limekiln Point on San Juan Island.

**buoyant uplift.** Crustal rocks uplifted by a floating equilibrium in the plastic asthenosphere when a load on the earth's surface is diminished, as when continental glaciers melt and mountains are eroded; the rigid rock above floats somewhat like a block of wood in water.

**breccia.** Fragmental rock made of relatively coarse (> 1/10 in.), angular chunks. Igneous breccia forms by explosive volcanic action. Sedimentary breccia is typically forms from talus at the base of a cliff. Fault breccia forms by breakage in a fault zone.

**carbon 14.** Isotope of carbon that decays to nitrogen, and is thus useful for radiometric dating

**calcite.** Carbonate mineral ($CaCO_3$). For our purposes, this is the mineral that makes up limestone and is derived from fossil shells; change to the polymorph aragonite occurs with deep burial and metamorphism.

**chemical precipitation.** Process of crystallization of minerals from a saturated solution; could be a saline lake, ground water, or igneous heated water (Yellowstone).

**chemical weathering.** Breakdown of minerals on the earth's surface primarily by dissolution in surface waters. Quartz is extremely resistant, all other common minerals are susceptible.

**chert.** Rock made of fine-grained quartz or opal. In the San Juan Islands, chert comes from accumulations of siliceous plankton – radiolaria.

**clastic.** Textural term applied to sedimentary rock that is made of fragments, for example: shale, sandstone, conglomerate, breccia.

**chlorite.** Green platy metamorphic mineral, altered from igneous pyroxene or amphibole. Iron-magnesium-aluminum silicate.

**cleavage.** Planar fabric that penetrates a metamorphic rock – like pages in a book. Is caused by squeezing of the rock to align platy minerals, such as mica.

**Cocos Plate.** Converges against and is subducted under Mexico and Central America.

**confining pressure.** Pressure coming equally from all sides, related to depth of burial.

**conglomerate.** Clastic sedimentary rock made up of pebbles and cobbles that are at least somewhat rounded, distinguishing conglomerate from breccia.

**conodont.** Fossil jaw bone or teeth of an extinct eel-like creature, useful for age analysis of Paleozoic rocks.

**continental arc.** Belt of igneous rocks developed in a continental land mass, over a subduction zone.

**cross-bedding.** Beds of sand or gravel inclined to the prevailing bedding orientation. Typically formed in river deposits as a sandbar at a bend in the river.

**crystal lattice.** Three dimensional configuration of atoms (or ions) bonded together constituting the internal structure of a crystal.

**cumulate gabbro.** Coarse-grained dark igneous rock which formed by crystallization, settling, and accumulation of mineral grains on the bottom of the magma chamber.

**dacite.** Light colored volcanic rock. Similar to rhyolite but lacks apparent quartz.

**decay rate.** Time it takes for a radioactive substance to transform into a daughter product; for example uranium giving off radiation and changing to lead. Usually expressed as the half-life.

**detrital zircon.** Zircon occurring as sedimentary grains.

**derivative magma.** Magma, molten rock, formed by partitioning of an original magma into: parts that have crystallized, and remaining magma that is changed in composition from the original by subtraction of the crystallized parts.

**dikes.** Sheet-like intrusions of igneous rock intruded across structure in the country-rock.

**diorite.** Coarse-grained igneous rock, not as dark as gabbro, usually has visible grains of light-colored feldspar, and dark amphibole.

**ductile deformation.** Process of rock deformation by bending or flowing, not breaking.

**earth's core.** Innermost part of the earth; made of an iron-nickel alloy.

**earth's mantle.** Zone in the earth between the core and crust, about 1800 miles thick, made of magnesium-iron silicate minerals.

**electron microscope.** Microscope that shines electrons on the sample of interest to obtain an image resolving detail on a very fine scale.

**epidote.** Metamorphic mineral distinguished by its pistachio green color. Calcium-aluminum silicate.

**erratic.** Rock moved and deposited by a glacier.

**eustatic.** Sea level fluctuation that is related only to the amount of water in the sea, not land movements.

**extensional basin.** Sedimentary basin formed by the pulling apart of the sides, and dropping of the floor.

**Farallon Plate.** Ancient crustal plate of the ancestral Pacific Ocean. Most of the plate has been subducted along the western margin of North and South America. Remnants are the Cocos, Nazca, and Juan de Fuca plates.

**fault.** Break in rocks where one side has slide relative to the other side.

**Fe.** Iron.

**feldspar.** Light-colored aluminum-silicate mineral common in igneous rocks, and in arkosic sandstone.

**fold.** Bent rock. Folds can be tight or open, minute or mountain size.

**foram fossils.** Short for foraminifera. In rocks of the San Juan Islands, these are small bottom-living marine animal with an external bean-shaped shell made of calcite.

**forearc.** Geologic setting in front of the arc, on the side of the subducting plate.

**foliation.** Planar fabric in metamorphic rocks caused by alignment of platy minerals, especially mica.

**fractionation of magma.** Process of magma freezing to crystallizes different minerals at different stages of cooling. Early-formed minerals are separated from remaining magma by, for example, intruding the remaining magma as a dike into the countryrock, then the original magma has fractionated into different rock types.

**fusulinids.** Type of foram fossil, very useful for late Paleozoic analysis of age and geographic origins.

**graded bedding.** Clastic sedimentary beds that grade from coarse to fine going upward, as a consequence of settling of grains from a turbid slurry of sediment and water.

**granite.** Coarse-grained light-colored igneous rock, made mostly of quartz and feldspar.

**geochronology.** Science of determining rock ages.

**geochronometer.** Kind of tool for determining rock ages, for example radioactive decay of uranium to lead.

**geothermal gradient.** Rate of increase of temperature with depth in the earth.

**glaciomarine drift.** Glacier-carried material deposited below sea level.

**greenstone.** Fine-grained green metamorphic rock.

**greywacke.** Sandstone comprised of dark sand grains of mainly volcanic rock and chert. Mud matrix is common.

**heavy liquid.** Chemical with greater density than most minerals, useful for separating heavier from lighter minerals

**hornblende.** Type of amphibole. Iron-magnesium silicate. Dark, elongate, found in diorite.

**hot spot.** Localized zone of hot upwelling of mantle material that causes melting to produce basalt. Good example is the Hawaiian hot spot.

**igneous rock.** Rock that forms by cooling and solidification of melt (magma).

**injection complex.** Multiple dike intrusions commonly observed in a single outcrop.

**island arc.** Chain of oceanic volcanic islands developed over a subduction zone.

**isostatic.** In the San Juan Islands, this term refers to the buoyant uplift of land when the glaciers melted; the crust floated up as the load on the underlying asthenosphere was reduced.

**isotope.** Many chemical elements have versions with different make-ups of the atom in terms of number of neutrons. Thus, carbon 12 has 6 protons and 6 neutrons; whereas carbon 14 has the same number of protons, but 8 neutrons. Carbon 12 is stable; carbon 14 changes (decays) to nitrogen.

**Juan de Fuca Plate.** Small remnant off the coast of the Pacific Northwest of the ancestral Farallon Plate. The Juan de Fuca Plate is actively subducting and gives rise to the Cascade volcanoes.

**Laurentia.** Ancestral North America in Precambrian times, before the addition of all rocks west of the Rocky Mountains.

**layered gabbro.** Light and dark layers in gabbro formed by settling of pulses of feldspar vs. pyroxene crystals onto the floor of the magma chamber.
**limestone.** Sedimentary rock made of calcite, generally derived from fossil shells.
**lithification.** Process of formation of sedimentary rock from loose sediment, with hardening by compaction and cementation.
**lithosphere.** An outer part of the earth consisting of the uppermost mantle and earth's crust.
**Ma.** Millions of years ago.
**magnetic anomaly.** Place on earth where the magnetic field (measured by a magnetometer!) is different from normal for the area.
**magma.** Molten rock.
**magma chamber.** Large blob inside the earth, in the crust or mantle, where magma has accumulated.
**mass spectrometer.** Instrument that distinguishes particles of different mass. Can measure different isotopes and is thus critical for some types of isotopic age-dating, e.g. uranium-lead.
**mechanical weathering.** Breakdown of rocks by mechanical means, as opposed to chemical weathering. Examples are frost wedging, rock fracture by falling off a cliff or bouncing in a stream.
**metamorphic rock.** Rock forms when a sedimentary or igneous rock is buried in the earth or intruded by a pluton so that increased heat and pressure change the minerals and rock textures.
**mineral stability.** An unstable mineral can dissolve in water or change to another mineral, depending on the temperature, pressure and composition of the water. A stable mineral has no force upon it to change.
**Mg.** Magnesium
**mica.** A platy mineral with well developed cleavage that allows thin sheets to be pealed off.
**moho.** Boundary between the earth's crust and underlying mantle.
**mudstone.** Dense, fine-grained sedimentary rock formed by compaction of mud.
**nappes.** Large, miles wide, sheets of rock that have been thrust over the landscape.
**Nazca Plate.** Remnant of the ancestral Farallon Plate that is currently being subducted under the west coast of South America
**nitrogen 14.** Isotope of nitrogen that forms by decay of carbon 14.
**North American Cordillera.** Tract of land extending approximately from the Rock Mountains to the west coast, including sedimentary rocks along the western margin of Laurentia and much other rock added to the continent by way of terrane accretion, plutonism and vulcanism.
**ocean crust.** Consists of ocean floor basalt generated at an ocean ridge system or an ocean island hot spot. Older crust has a sediment overlay, typically of chert.
**ocean island.** Volcanic island generated over a *hot spot* within the oceanic plate. A long-lasting hot spot will create a chain of islands as the ocean crust moves along over the magma source (Hawaiian chain).
**ocean ridge.** Ridge system formed where ocean plates move apart. Basalt magma is generated in the upper mantle, intrudes upward to the ridge where it crystallizes, adds to the ocean crust.
**ocean trench.** Large-scale depression formed where the ocean plate dives down into a subduction zone. The trench is coupled with an arc formed over the subduction zone.
**olivine.** Magnesium silicate mineral that is a major component of the upper mantle; occurs in the rock peridotite.
**ophiolite.** An on-land sequence of rocks that is an uplifted oceanic formation; some ophiolites are crustal sections of an island arc, other ophiolites represent ocean crust as generated at the ridge system.
**outcrop.** Exposure of rock.
**outwash.** Sediment washed out of a glacier by meltwater streams.
**passive margin.** Continental margin not affected by plate collision; no volcanic arcs, limited earthquakes.
**paleontology.** Study of fossils
**pelagic.** Open ocean environment, not on the sea floor.
**pelagic argillite.** Strongly compacted mudstone derived from mud carried in the open ocean and deposited on the sea floor.
**peridotite.** Rock made of olivine and pyroxene, prevalent in the upper mantle.
**pillow basalt.** Basalt formed with pillow structure as a consequence of eruption under water.
**placer-mined.** Heavy minerals concentrated from a mixture of minerals by swirling away of the lighter minerals in turbulent water, as in gold panning.
**planar fabric.** Planes of alignment of platy minerals, typically in a metamorphic rock. Foliation.
**plankton.** Pelagic organisms that drift with ocean currents.
**plastic.** Condition of a rock where it can flow as a solid.
**plate tectonics.** Theory of dynamic mechanisms in the outer earth. The earth, from the crust into the upper mantle, is divided into plates, each of which is up to tens of miles thick and thousands of miles in breadth. The plates slide about on a slippery substrate and their motions cause earthquakes, volcanoes, and virtually all other large-scale movements of the earth's crust.
**pluton.** Igneous rock that is coarse grained and therefore crystallized slowly well below the earth's surface.
**polymorph.** A solid of a certain chemical formula can take different structural forms. Calcite and aragonite, both $CaCO_3$, have different internal structure.
**prehnite.** Green metamorphic mineral. Calcium-aluminum silicate.

**provenance.** Source area, as for example, the rock from which sand grains were eroded.
**pumpellyite.** Green metamorphic mineral. Calcium-iron-magnesium-aluminum silicate.
**pyroclastic.** Fragmental texture in an igneous rock produced by an explosive volcanic eruption.
**pyroxene.** Iron-magnesium silicate mineral common in gabbro and peridotite.
**quartz.** Silicon dioxide, common in rocks, resistant to weathering.
**radioactivity.** Emission of radiation caused by breakup of the nucleous of an atom. Alpha particles, beta particles and gamma rays are emitted. Dangerous near concentrations of uranium.
**radiometric age.** Age of a mineral determined by analyzing parent and daughter products of radioactive decay.
**radiolaria.** Tiny, siliceous, planktonic organisms. Mass accumulation of dead radiolaria raining down on the sea floor is the raw material for creating chert.
**remnant arc.** Island arc that is no longer an active volcano, having lost its connection to the subduction zone magma source as a consequence of the subducting plate stepping back, away from the arc.
**ribbon chert.** Chert beds interlayered with softer, recessive shale beds.
**Rodinia.** At different times in earth history all the continents have been joined together in one giant landmass. Rodinia existed about from about 1.1 to 0.75 billion years ago, then split into numerous continents, including Laurentia, the ancestral North America. The continents all got back together at about 300 million years ago, forming the supercontinent Pangea, which split about 200 million years ago to yield early forms of the continents we have now.
**seamount.** Extinct ocean island volcano that due to plate motion has moved off its hot spot magma source.
**sea-floor spreading.** Spreading of oceanic plates away from an ocean ridge system. The plate grows by addition of basalt at the ridge, then moves away, downhill, toward a subduction zone.
**sedimentary rock.** Rock formed at the earth's surface by accumulation and hardening of various substances: weathered material (sand etc.), organisms (plant or animal), or by chemical precipitates (salt or other).
**seismic.** Having to do with earthquakes.
**seismic waves.** Earthquake energy is propagated by waves of earth motion.
**serpentinite.** Rock made of the mineral serpentine, a magnesium rich silicate that forms by alteration of olivine in peridotite.
**shear pressure.** Directed stress on a rock causing a displacement by sliding of one part past another, in contrast to confining pressure which squeezes a rock in all directions, making it smaller.
**siliceous.** Rich in silica.
**sills.** Sheets of igneous rock intruded parallel to structure in the country rock (e.g. sedimentary bedding).
**sorting.** Process that occurs where clastic sediment of mixed sizes is water-washed or windblown to separate grains according to how much energy is needed to move them.
**subaerial.** On the earth's surface under the air.
**strike and dip.** Measurements that give the orientation of a planar geologic surface, such as sedimentary bedding. The strike is the compass direction of a horizontal line lying in the plane. The dip line is at right angles to the strike, it follows the fall-line down the surface, and the dip reading is the angle between the dip line and horizontal.
**subduction zone.** Region where ocean crust collides with another plate and plunges down into the mantle.
**tectonics.** Relates to large-scale crustal movements: plate interactions, mountain building etc.
**terrane.** Large crustal block of rock that has moved from a place of origin to be faulted into where we find it now. This rock is "exotic" to the land in which it is found.
**Tethys Sea** Part of the ancestral Pacific Ocean between Asia and eastern Africa about 250-150 million years ago.
**thrust sheet.** Nearly horizontal slab of rock pushed over or under a backstop of land.
**tuff.** Volcanic ash.
**turbidites.** Clastic sedimentary rock deposited in the ocean (or lake) from a turbid slurry of mud, silt and sand. Beds are typically graded upward from coarse to fine grains.
**under-plated.** Rock materials added to the continental margin by sticking on of thrust sheets to the bottom of the over-riding plate in a subduction zone.
**volcanic rock.** Erupted igneous rock.
**weathering.** Process of natural breakdown of rocks and minerals. Chemical weathering occurs by dissolution in surface waters; mechanical weathering is fracturing caused by frost wedging, transport over cliffs, bouncing in river beds, pounding in surf zones, etc.
**Wrangellia.** Large terrane faulted onto the outermost northwest margin of North America, extending from Vancouver Island to Alaska. The terrane has a long history of formation as ocean crust, 350-150 Ma, including development of oceanic plateaus and island arcs.
**x-ray diffraction.** Interaction of x-rays with crystal structure. The pattern of diffracted x-rays allows identification of the mineral.
**zircon.** Zirconium silicate. A common, but very minor, mineral in igneous rocks and sediments eroded from igneous rocks. Trace amounts of uranium in zircon allow radiometric age determination.

# REFERENCES

**Introduction.** *33*
**Chapter 3. Rock terminology and origins.** *15, 17, 47*
**Chapter 4. Overview of the San Juan Islands.** *5, 8, 9, 23, 33, 36, 37, 38, 46, 48, 49, 50, 53*
**Chapters 5 and 6. Turtleback Complex and East Sound Group.** *5, 11, 18, 20, 48, 49*
**Chapter 7. Orcas Chert and Deadman Bay Volcanics.** *5, 19, 20, 42, 48, 49*
**Chapter 8. Ocean Floor Assemblage.** *2, 3, 10, 13, 14, 45, 48, 49*
**Chapter 9. Fidalgo Ophiolite.** *6, 12, 27, 50, 54*
**Chapter 10. Nanaimo Group.** *8, 28, 32, 35, 39, 51*
**Chapter 11. Chuckanut Formation.** *30, 40, 41*
**Chapter 12. Terrane travels.** *2, 7, 9, 16, 21, 25, 29, 31, 34, 44, 50*
**Chapter 13. Continental Glaciation.** *1, 4, 22, 24, 26, 43, 52*

1. Armstrong, J. E., Crandell, D. R., Easterbrook, D. J., and Noble, J. B. 1965. Late Pleistocene stratigraphy and chronology in southwestern British Columbia and northwestern Washington. Geol. Society of America Abstracts Bulletin, 76: 321 - 330.
2. Bergh, S. G. 2002. Linked thrust and strike-slip faulting during Late Cretaceous terrane accretion in the San Juan thrust system, Northwest Cascade orogen, Washington. Geological Society of America Bull., 114: 934-949.
3. Blake, M. C., Jr., Burmester, D. C., Engebretson, D. E., and Aitchison, J. 2000. Accreted terranes of the eastern San Juan Islands, Washington. Geological Society of America Abstracts with Programs, 32: 4.
4. Booth, D. B., Troost, K. G., Clague, J. J., and WaitR. B. 2003. The Cordilleran Ice Sheet. *In* Development in Quaternary Science. *Edited by* A. R. Gillespie, Porte, S.C., Atwater, B.F. Elsevier, one, pp. 17-43.
5. Brandon, M. T., Cowan, D. S., and Vance, J. A. 1988. The Late Cretaceous San Juan thrust system, San Juan Islands, Washington. Geological Society of America Special Paper 221, 81 p.
6. Brown, E. H. 1977. Ophiolite on Fidalgo Island, Washington. *In* North American Ophiolites. *Edited by* R. G. Coleman and W. P. Irwin. Oregon Department of Geology and Mineral Industries, Bulletin 95, pp. 67-93.
7. Brown, E. H. 1987. Structural geology and accretionary history of the Northwest Cascades System, Washington and British Columbia. Geological Society of America Bulletin, 99: 201-214.
8. Brown, E. H. 2012. Obducted nappe sequence in the San Juan Islands - northwest Cascades thrust system, Washington and British Columbia. Canadian Journal of Earth Sciences, 49: 796-819.
9. Brown, E. H., and Dragovich, J. D. 2003. Tectonic elements and evolution of northwest Washington. Washington Division of Geology and Earth Resources Geologic Map GM-52:
10. Brown, E. H., and Gehrels, G. E. 2007. Detrital zircon constraints on terrane ages and affinities and timing of orogenic events in the San Juan Islands and North Cascades, Washington. Canadian Journal of Earth Sciences, 44: 1375-1396.
11. Brown, E. H., Gehrels, G. E., and Valencia, V. A. 2010. Chilliwack composite terrane in northwest Washington: Neoproterozoic-Silurian passive margin basement, Ordovician-Silurian arc inception. Canadian Journal of Earth Sciences, 47: 1347-1366.
12. Brown, E. H., Housen, B. A., and Schermer, E. R. 2007. Tectonic evolution of the San Juan Islands thrust system. *In* Floods, Faults Society of America, Boulder, Field Guide 9, pp. 143-177.
13. Brown, E. H., Lapen, T. J., Leckie, R. M., Premoli Silva, I., Verge, D., and Singer, B. S. 2005. Revised ages of blueschist metamorphism and the youngest pre-thrusting rocks in the San Juan Islands, Washington. Canadian Journal of Earth Sciences, 42: 1389-1400.
14. Burmester, R. F., Blake, M. C., Jr., and Engebretson, D. C. 2000. Remagnetization during Cretaceous Normal Superchron in Eastern San Juan Islands, WA: implications for tectonic history. Tectonophysics, 326: 73-92.
15. Carlson, W. D., and Rosenfeld, J. L. 1981. Optical determination of topotactic aragonite-calcite growth kinematics: metamorphic implications. Journal of Geology, 89: 615-638.
16. Colpron, M., and Nelson, J. A. 2009. A Paleozoic Northwest Passage: Incursion of Caledonian, Baltican and Siberian terranes into eastern Panthalassa, and the early evolution of the North American Cordillera. *In* Accretionary Orogens through Space and Time, *Edited by* P. C. A. Kröner. Geol. Society of London Special Publication, Special Publications, v. 318, pp. 273-307.
17. Crawford, W. A., and Hoersch, A. L. 1972. Calcite - aragonite equilibrium from 50 degrees C to 150 degrees C. American Mineralogist, 57: 995-998.
18. Danner, W. R. 1966. Limestone resources of western Washington, with a section on the Lime Mountain deposit by G.W. Thorsen. Washington Division of Mines and Geology Bulletin 52, 474 p.
19. Danner, W.R. 1970. Paleontologic and stratigraphic evidence for and against sea floor spreading and opening and closing oceans in the Pacific Northwest. Geological Society of America Abstracts with programs, v.2, no. 2, 84-85.
20. Danner, W. R. 1977. Paleozoic rocks of northwest Washington and adjacent parts of British Columbia. In Stewart, J.H., Stevens, C.H., Fritsche, A.E., editors, Paleozoic paleogeography of the western United States: Society of Economic Paleontologists and Mineralogists Pacific Section, Pacific Coast Paleogeography Symposium 1, p. 481-502.
21. Davis, G. A., Monger, J. W. H., and Burchfiel, B. C. 1978. Mesozoic construction of the Cordilleran "collage", central British Columbia to central California. *In* Mesozoic Paleogeography of the Western United States, Proceedings of the Pacific Coast Paleogeography Symposium. *Edited by* D. G. Howell and K. A. McDougal. Society of Economic Paleontologists and Mineralogists, Pacific Section, Los Angeles, California, 2, pp. 1-32.
22. Dethier, D. P., Pessl,, Fred, Keuler, R. F., Balzarini, M. A., and Pevear, D. R. 1995. Late Wisconsin glaciomarine deposition and isostatic rebound, northern Puget Lowland, Washington. Geological Society of America Bulletin, 107: 1288 - 1303.

23. Dragovich, J. D., Logan, R. L., Schasse, H. W., Walsh, T. J., Lingley, W. S., Norman, D. K., Gerstel, W. J., Lapen, T. J., Schuster, J. E., and Meyers, K. D. 2002. Geologic Map of Washington - Northwest Quadrant. Washington Division of Geology and Earth Resources, Geologic Map GM-50.
24. Easterbrook, D. J. 1969. Pleistocene Chronology of the Puget Lowland and San Juan Islands. Geol. Soc. Amer. Bull., 80: 2273 - 2285.
25. Enkin, R. J., Baker, J., and Mustard, P. S. 2001. Paleomagnetism of the Upper Cretaceous Nanaimo Group, southwestern Canadian Cordillera. Canadian Journal of Earth Sciences, 38: 1403-1422.
26. Fleming, K., Johnson, P., Zwartz, D., Yokoyama, Y., Lambeck, K., and Chappell, J. 1998. Refining the eustatic sea-level curve since the Last Glacial Maximum using far- and intermediate-field sites. Earth and Planetary Science Letters, 163: 327-342.
27. Gusey, D., and Brown, E. H. 1987. The Fidalgo ophiolite, Washington. *In* Cordilleran section of the Geological Society of America Centennial Field Guide. *Edited by* M. L. Hill. Geological Society of America, Boulder, CO, 1, pp. 389-392.
28. Haggart, J. W., Ward, P. D., and Orr, W. 2005. Turonian (Upper Cretaceous) lithostratigraphy and biochronology, southern Gulf Islands, British Columbia, and northern San Juan Islands, Washington State. Can. Journal of Earth Sciences, 42: 2001- 2020.
29. Irving, E., Woodsworth, G. J., Wynne, P. J., and Morrison, A. 1985. Paleomagnetic evidence for displacement from the south of the Coast Plutonic Complex, British Columbia. Canadian Journal of Earth Science, 22: 584-598.
30. Johnson, S. Y. 1984. Stratigraphy, age, and paleogeography of the Eocene Chuckanut Formation, northwest Washington. Canadian Journal of Earth Sciences, 21: 92-106.
31. Kodama, K. P. and Ward, P. D. 2001. Compaction-corrected paleomagnetic paleolatitudes for Late Cretaceous rudists along the Cretaceous California margin; evidence for less than 1500 km of post-Late Cretaceous offset for Baja British Columbia. Geological Society of America Bulletin, 113: 1171-1178.
32. Mahoney, J. B., Link, P. K., Mustard, P. S., and Fanning, C. M. 2003. Belt Supergroup detritus in the Nanaimo Group; a robust provenance tie to northern latitudes. Geological Society of America Abstracts with Programs, 35: 390.
33. McClellan, R. D. 1927. Geology of the San Juan Islands (Washington). PhD thesis, University of Washington, Seattle.
34. McGroder, M. F. 1991. Reconciliation of two-sided thrusting, burial metamorphism, and diachronous uplift in the Cascades of Washington and British Columbia. Geological Society of America Bulletin, 103: 189-209.
35. Mercier, J.M. 1977. Petrology of the Upper Cretaceous Strata of Stuart Island, San Juan County, Washington. PhD thesis, Washington State University.
36. Miller, M.M. 1987. Dispersed remnants of a northeast Pacific fringing arc: upper Paleozoic terranes of Permian McCloud faunal affinity, western U.S. Tectonics,6:807-830.
37. Miller, R. B. 1985. The ophiolitic Ingalls Complex, north-central Cascade Mountains, Washington. Geol. Soc. of Amer. Bull, 96 27-42.
38. Misch, P. 1966. Tectonic evolution of the northern Cascades of Washington State - a west-Cordilleran case history. *In* A symposium on the tectonic history and mineral deposits of the western Cordillera in British Columbia and neighboring parts of the United States. *Edited by* H. C. Gunning. Canadian Institute of Mining and Metal, Vancouver, B.C., 1964, Special Volume 8: 101-148.
39. Mustard, P. S. 1994. The Upper Cretaceous Nanaimo Group, Georgia Basin. *In* Geology and Geologic Hazards of the Vancouver Region, southwestern British Columbia. *Edited by* J. W. H. Monger. Geological Survey of Canada Bulletin 481, pp. 27-95.
40. Mustoe, G. E. 2010. Biogenic origin of coastal honeycomb weathering. Earth surfaces processes and landforms, (www.interscience.wiley.com) DOI: 10.1002/esp.1931.
41. Mustoe, G. E., Dillhoff, R. M., and Dillhoff, T. A. 2007. Geology and paleontology of the early Tertiary Chuckanut Formation. *In* Floods, Faults, and Fire: Geological Field Trips in Washington State and Southwest British Columbia. Geological Society of America Field Guide 9. *Edited by* P. Stelling and D. S. Tucker. pp. 121 - 135.
42. Ota, Y., Yamagata, T., and Danner, W. R. 2009. *Yabeina cascadensis* (Anderson) (Permian Fusulinacea) from San Juan Island, Washington, USA. Regional Views; Institute for Applied Geography, Komazawa University, Tokyo, 22: 17-45.
43. Porter, S. C., and Swanson, T. W. 1998. Radiocarbon Age constraints on rates of advance and retreat of the Puget Lobe of the Cordilleran Ice Sheet during the last glaciation. Quaternary Research, 50: 205 - 213.
44. Sauer, K. et al. 2014. Tectonic implications of detrital zircon geochronology and neodymium isotopes of the arkosic petrofacies of the Western Mélange Belt, Lake Champlain Quadrangle, Western Cascades, Washington. Geological Society of Americas Abstracts with Programs. Vol.46, No. 6, p.363.
45. Schermer, E. R., Gilaspy, J. R., and Lamb, R. 2007. Arc-parallel extension and fluid flow in an ancient accretionary wedge: the San Juan Islands, Washington. Geological Society of America Bulletin, 119: 753-767.
46. Tabor, R. W., Haugerud, R. A., Hildreth, W., and Brown, E. H. 2003. Geologic Map of the Mt Baker 30- by 60 Minute Quadrangle, Washington. U.S. Geological Survey Miscellaneous Investigations Map I-2660, scale 1:100,000.
47. Vance, J. A. 1968. Metamorphic aragonite in the prehnite-pumpellyite facies, northwest Washington. Amer. Jour. of Sci., 266: 299-315.
48. Vance, J. A. 1975. Bedrock geology of San Juan County. *In* Geology and water resources of the San Juan Islands, San Juan County, Washington. *Edited by* R. H. Russell. Washington Department of Ecology, Water-Supply Bulletin 46, pp. 3-19.
49. Vance, J. A. 1977. The stratigraphy and structure of Orcas Island, San Juan Islands. *In* Geological Excursions in the Pacific Northwest. *Edited by* E. H. Brown and R. C. Ellis. Western Washington University, Bellingham, Wash., pp. 177-203.
50. Vance, J. A., Dungan, M. A., Blanchard, D. P., and Rhodes, J. M. 1980. Tectonic setting and trace element geochemistry of Mesozoic ophiolitic rocks in Western Washington. American Journal of Science, 280-A: 359 - 388.
51. Ward, P. D. 1978. Revisions to the stratigraphy and biochronology of the Upper Cretaceous Nanaimo Group, British Columbia and Washington State. Canadian Journal of Earth Sciences, 15: 405-423.
52. Whetten, J. T. 1975. The geology of the southeastern San Juan Islands. *In* Geology and water resources of the San Juan Islands, San Juan County Washington. *Edited by* R. H. Russell. Washington Department of Ecology, Water-Supply Bulletin 46, pp. 41-57.
53. Whetten, J. T., Jones, D. L., Cowan, D. S., and Zartman, R. E. 1978. Ages of Mesozoic terranes in the San Juan Islands, Washington. *In* Mesozoic paleogeography of the western United States, Proceedings of the Pacific Coast Paleogeography Symposium: *Edited by* D. G. Howell and K. A. McDougal. Soc. of Econ. Paleontol. and Mineralogists Pacific Section, Vol. 2, pp. 117-132.
54. Whetten, J. T., Zartman, R. E., Blakely, R. J., and Jones, D. L. 1980. Allochthonous Jurassic ophiolite in Northwest Washington. Geological Society of America Bulletin, 91: 359-368.